Building services and equipment

Volume 3

Building services and equipment

Volume 3
Third edition

Fred Hall
H.N.C., C.Ed., C.G.F.T.C., Dip C.I.O.B., A.I.P.

Routledge
Taylor & Francis Group
LONDON AND NEW YORK

First published 1980 by Pearson Education Limited
Second edition 1986
Third edition 1994

Published 2014 by Routledge
2 Park Square, Milton Park, Abingdon, Oxon OX14 4RN
711 Third Avenue, New York, NY 10017, USA

Routledge is an imprint of the Taylor & Francis Group, an informa business

ISBN 13: 978-0-582-23139-9 (pbk)

British Library Cataloguing in Publication Data
A catalogue entry for this title is available from the British Library.

Set by 18 in 10/12pt IBM Journal

Contents

Preface

The Business Technology Education Council (BTEC) standard unit in Building Services and Equipment for the award of the higher diploma or certificate, requires the student to develop an appreciation of the principles of services design in various specific areas.

This volume has been specially written for students studying for the standard module and the text covers closely the requirements of the BTEC syllabus.

Students in building, architecture and surveying studying for the higher diploma or certificate should therefore find this volume a useful addition to *Building Services and Equipment*, volumes 1 and 2. This book should also be of assistance to students studying for the Chartered Institute of Building and the Royal Institution of Chartered Surveyors' examinations in services.

The text covers, by carefully selected worked examples the principles of services design and the questions with answers at the end of each chapter should help the student to obtain a better understanding of the topics covered. While primarily intended for the srudent attending lectures, the book is also designed so that the reader will have little difficulty in studying the subjects covered by private study.

Practising technicians in building, architecture and surveying should also find the book useful by providing an appreciation of the design of the building services.

SI units are used throughout the book and a description of the units has been included. Wherever possible, diagrams and tables have been included to supplement the written descriptions and calculations.

I should like to thank the publishers, C. R. Bassett for his encouragement and helpful constructive criticisms, and my late wife for typing the manuscript, without whose help this book would never have been written.

F.Hall
1994

Acknowledgements

The author is grateful to the following for permission to reproduce copyright material:
The Chartered Institution of Building Services; the Illuminating Engineering Society and the Wednesbury Tube Company.

Chapter 1

Speed, velocity and acceleration

Speed, velocity and acceleration

Speed is defined as the rate of change of distance moved with time. Velocity is defined as the rate of change of distance moved with time in a specified direction. Speed is a scalar quantity whereas velocity is a vector.

When travelling by car it is speed that can be read on the speedometer because the direction is unimportant. A speed of 50 km/h is known as a scalar quantity while a velocity of 50 km/h due south is a vector. Velocity can therefore be represented by a straight line drawn to scale in a given direction.

Average speed

Unless travelling on a motorway, a motorist cannot usually maintain a constant speed for any length of time. When deciding the time to allow for a particular journey, the motorist must have some idea of the average speed at which it is possible to travel.

Example 1.1. *Find the average speed during a journey of 276 km which takes 6 hours.*

$$\text{average speed} = \frac{\text{distance}}{\text{time}} = \frac{276}{6} = 46 \text{ km/h}$$

Uniform velocity

A body is said to move with uniform velocity if its rate of change of distance with time in a specified direction is constant.

In services the distance and time are expressed in metres and seconds respectively.

$$\text{velocity} = \frac{\text{distance}}{\text{time}} = \frac{\text{metres}}{\text{seconds}} = \text{m/s}$$

Example 1.2. *A travelator (moving pavement) moves at a velocity of 0.6 m/s. Calculate the distance a person would move on the travelator in 30 seconds.*

$$\text{velocity} = \frac{\text{distance}}{\text{time}}$$

$$\begin{aligned}\text{distance} &= \text{time} \times \text{velocity}\\ &= 30 \times 0.6\\ &= 18 \text{ m}\end{aligned}$$

Acceleration

Acceleration is a natural progression from velocity, and is defined as the rate of change of velocity with time.

$$\text{acceleration} = \frac{\text{velocity}}{\text{time}} = \text{m/s}^2$$

Example 1.3. *A lift car accelerates uniformly from rest to 2 m/s in 4 seconds. Calculate the acceleration and the distance travelled by the car.*

$$\begin{aligned}\text{acceleration} &= \frac{\text{change in velocity}}{\text{time}}\\ &= \frac{2 \text{ m/s}}{4 \text{ s}}\\ &= 0.5 \text{ m/s}^2\end{aligned}$$

In order to calculate the distance travelled by the car, the average velocity is required.

$$\frac{\text{average velocity for}}{\text{uniform acceleration}} = \frac{\text{initial velocity} + \text{final velocity}}{2}$$

$$\begin{aligned}\text{average velocity} &= \frac{0 + 2}{2}\\ &= 1 \text{ m/s}\\ \text{distance travelled} &= 1 \times 4\\ &= 4 \text{ m}\end{aligned}$$

Example 1.4. *A lift car starts from rest and is accelerated uniformly at the rate of 0.5 m/s² for 5 s. It then maintains a constant velocity for 10 s. The car is then uniformly retarded at the rate of 0.5 m/s² for 5 s. Find the maximum velocity of the car and the total distance it travels.*

Method 1. By calculation

u = initial velocity in m/s
a = acceleration in m/s^2

v = final or maximum velocity in m/s
t = time for which acceleration continues in seconds
s = distance travelled in m

To find the maximum velocity

$u = 0$ m/s
$a = 0.5$ m/s^2
$t = 5$ s
$v = 0 + 0.5 \times 5$
$= 2.5$ m/s   which is the maximum velocity

First stage. Distance travelled is the average velocity m/s multiplied by time in seconds.

$$s = \frac{2.5}{2} \times 5$$

$$s = 6.25 \text{ m}$$

Second stage. distance moved = velocity × time
= 2.5 × 10
= 25 m

Third stage. The distance travelled in the third stage will be the same as that in the first stage.

total distance travelled = 6.25 + 25 + 6.25
= 37.5 m

Method 2. Graphical solution

Draw the velocity–time graph (Fig. 1.1).

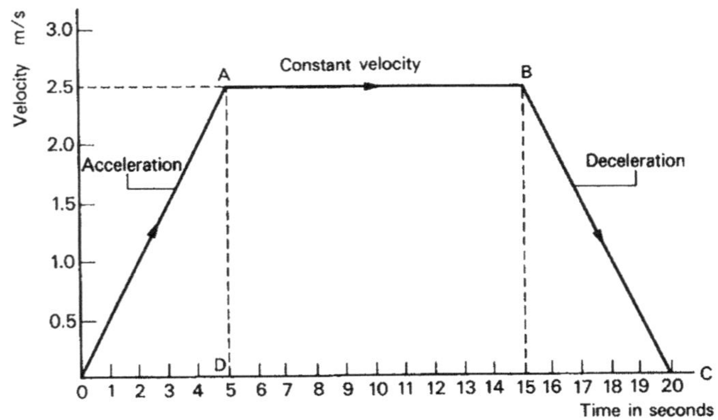

Fig.1.1 Velocity-time graph for a lift car

The area under the graph gives the distance travelled by the car.

$$\text{area OABC} = \frac{1}{2}(\text{AB} + \text{OC}) \times \text{AD}$$

$$= \frac{1}{2}(10 + 20) \times 2.5$$

$$= 37.5 \text{ m}$$

Gravitational acceleration

A well-known rate of acceleration is that due to gravity. If a body is falling freely near to the surface of the earth it has an acceleration due to gravity of 9.81 m/s^2, which is applicable to most geographical locations.

The velocity of a falling body does not depend upon mass but upon its nearness to the surface of the earth. In the study of dynamics in the sixteenth century, the Italian, Galileo, released simultaneously three iron balls of different masses from the top of the Leaning Tower of Pisa, and all three balls reached the ground at the same time. Before this time it was thought that the velocity of a falling body depended upon mass.

Example 1.5. *An iron ball is dropped from the top of a building and takes 3 seconds to reach the ground.*

(*a*) *What is its velocity when it hits the ground?*
(*b*) *What is the height of the building?*

terminal velocity = acceleration due to gravity × time
= 9.81 × 3
= 29.43 m/s

height of building = average velocity × time
(distance fallen)

$$= \frac{29.43}{2} \times 3$$

$$= 44.145 \text{ m}$$

It has been shown that the velocity of a falling body is not constant but increases uniformly during its period of fall. This increase in velocity or acceleration, as previously stated, is 9.81 metres/second for every second it is falling.

The velocity acquired by a body falling freely from rest for a period of t seconds would be

$$v = g\,t \qquad [1.1]$$

The initial velocity of a body falling from rest is zero and its velocity after t seconds is $g\,t$ metres/second.

The average velocity, since the increase is uniform, would be

$$\frac{\text{average velocity}}{\text{m/s}} = \frac{0 + g\,t}{2}$$

$$= \frac{g\,t}{2}$$

The distance fallen will be average velocity × time of fall

$$\text{distance fallen} = \frac{g\,t}{2} \times t$$
$$= \frac{g\,t^2}{2}$$

Substituting from equation [1.1]

$$t = \frac{v}{g}$$

$$h = \frac{g}{2} \times \frac{v}{g} \times \frac{v}{g}$$
$$= \frac{v^2}{2\,g}$$

Kinetic or velocity energy

The expression $h = v^2/2\,g$ gives the kinetic or velocity energy used in hydraulic calculations, and by transposing the formula, the theoretical velocity of water flowing through an orifice or a pipe may be found.

$$v = \sqrt{2\,g\,h} \qquad [1.2]$$

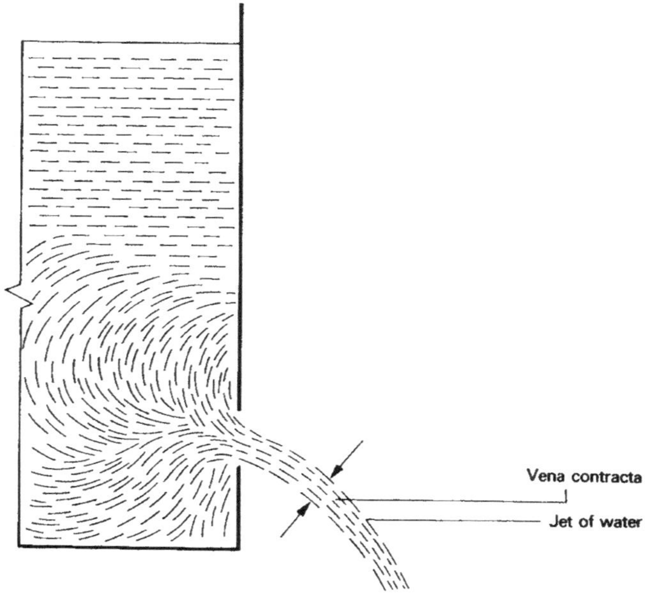

Fig.1.2 Flow of water through an orifice

where v = theoretical velocity of flow in m/s
g = 9.81 m/s^2
h = head of water in metres above the centre of the orifice or pipe

Due to the effect of friction, the theoretical velocity is not fully reached. When a jet of water is discharging freely through an orifice, the particles are trying to come together in all directions (see Fig. 1.2). This contraction is known as 'vena contracta' and is found by experiment to be about 0.64 of the orifice area. This is also termed the coefficient of contraction.

The effect of friction, known as the coefficient of velocity, is about 0.97 and the product of the coefficient of contraction and the coefficient of friction is known as the coefficient of discharge.

the coefficient of discharge $Cd = Cc \times Cv$

where Cc = coefficient of contraction
Cv = coefficient of velocity
$\therefore Cd = 0.64 \times 0.97 = 0.62$ (approx.)

The discharge through a pipe or orifice may be found from the formula

$$Q = VA$$

where Q = rate of discharge in m^3/s
V = velocity of flow in m/s
A = area of pipe or orifice in m^2

Substituting from equation [1.2]

$$Q = Cd \times A \times \sqrt{2\,g\,h}$$

Example 1.6. *Calculate the discharge in litres per second through a 100 mm diameter orifice at the base of a tank when there is a constant head of water above the centre of the orifice of 1.5 m.*

$$Q = Cd \times A \times \sqrt{2\,g\,h}$$
$$= 0.62 \times 3.142 \times 0.05 \times 0.05 \times \sqrt{2 \times 9.81 \times 1.5}$$
$$= 0.026\,4 \text{ m}^3/\text{s}$$

discharge = 26.4 litre/s

If a short length of pipe of the same diameter of the orifice is fixed to the tank, a different coefficient of discharge will result, depending upon the type of connection.

Figure 1.3 shows three different types of connections to the tank, and it will be seen that a bell-shaped connection will give the best discharge.

Example 1.7. *Calculate the time required to discharge 500 litres of water from a tank by means of a 50 mm diameter short pipe having a bell-shaped connection to the tank when there is a constant head of water above the centre of the pipe of 1.5 m.*

$$Q = Cd \times A \times \sqrt{2\,g\,h}$$
$$= 0.97 \times 3.142 \times 0.025 \times 0.025 \sqrt{2 \times 9.81 \times 1.5}$$
$$= 0.001\,905 \times 5.425$$
$$= 0.010\,335 \text{ m}^3/\text{s}$$

discharge = 10.335 litre/s

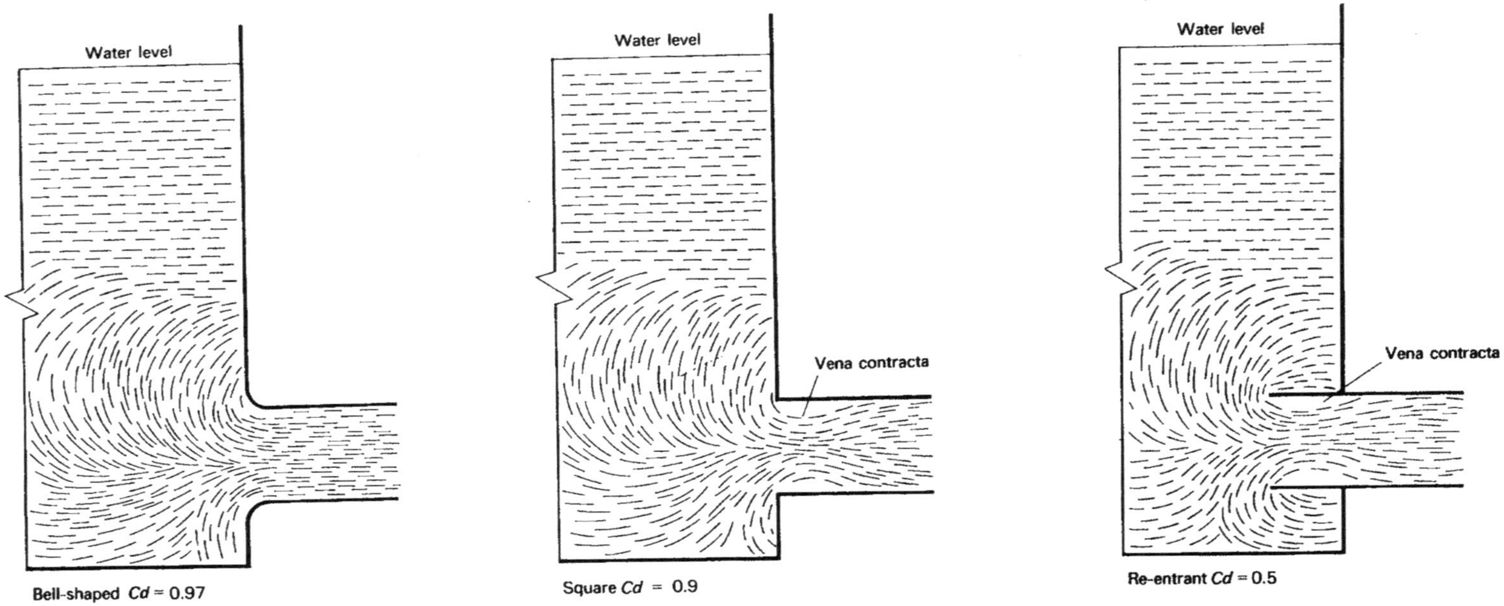

Fig.1.3 Pipe connections to tanks and cisterns

$$\text{time taken} = \frac{500}{10.335}$$
$$= 48.38 \text{ s}$$

The value of $Q = \pi R^2 \sqrt{2gh}$ will be constant for each case but the coefficient of discharge will of course vary.

$Q = 3.142 \times 0.05 \times 0.05 \times \sqrt{2 \times 9.81 \times 1.5}$
$= 0.007\,855 \times \sqrt{29.43}$
$= 0.042\,61 \text{ m}^3/\text{s}$ (theoretical)

Q (re-entrant outlet) $= 0.042\,61 \times 0.5$
$= 0.0213 \text{ m}^3/\text{s}$
$= 21.30$ litre/s

Q (square outlet) $= 0.042\,61 \times 0.9$
$= 0.038\,35 \text{ m}^3/\text{s}$
$= 38.35$ litre/s

Q (bell-shaped outlet) $= 0.042\,61 \times 0.97$
$= 0.041\,33 \text{ m}^3/\text{s}$
$= 41.33$ litre/s

Questions

1. A lift car starts from rest and is accelerated uniformly at the rate of 0.2 m/s^2 for 3 seconds. It then maintains a constant velocity for 8 seconds. The car then decelerates uniformly at the rate of 0.2 m/s^2 for 3 seconds. Find the maximum velocity of the car and the total distance it travels.

Answers: 0.6 m/s; 6.6 m

2. Draw a velocity–time graph for the movement of the lift car in question 1.
3. Define: (*a*) speed; (*b*) velocity; (*c*) acceleration; (*d*) gravitational acceleration.
4. An escalator moves at a velocity of 0.6 m/s. If the steps in use make up a distance of 8 m, calculate the time it would take a person to move up one floor.

Answer: 13.33 seconds

5. Calculate the velocity of water in a pipe of cross-sectional area of 0.02 m^2 which delivers 115 litres of water per second.

Answer: 5.75 m/s

6. Derive the formula for kinetic or velocity energy of a falling body.
7. Define: (*a*) vena contracta; (*b*) coefficient of discharge; (*c*) coefficient of velocity.
8. Compare the discharge of water through an orifice and the following short pipe connections to a tank: (*a*) re-entrant; (*b*) square; (*c*) bell-shaped.

9. Calculate the discharge in litres per second through a 50 mm diameter orifice at the side of a tank when there is a constant head of water above the centre of the orifice of 2 m.

Answer: 24.11 litre/s

10. Calculate the discharge in litres per second and the time required to discharge 300 000 litres of water from a swimming pool by means of a 76 mm diameter short pipe having a bell-shaped connection when there is a constant head of water above the centre of the pipe of 2 m.

Answers: 27.56 litre/s; 3 hours (approx.)

Chapter 2

Force and pressure

Force

The SI provides a coherent system of units in which the unit of force, the newton, can be used for calculations involving pressure, stress, work and power.

A force of 1 newton (N) is defined as that force which gives a body having a mass of 1 kilogram an acceleration of 1 metre per second squared (see Fig. 2.1).

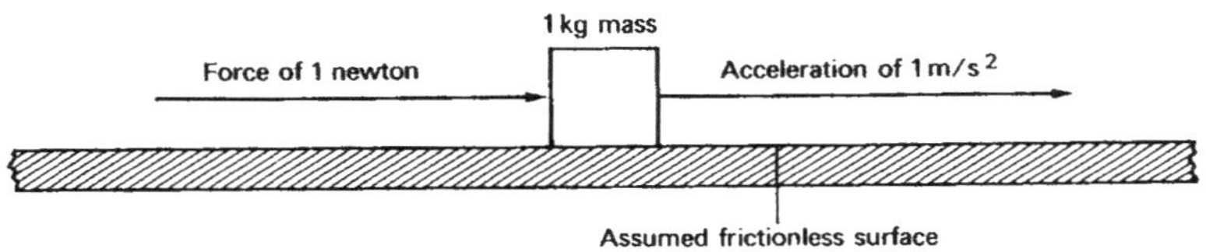

Fig.2.1 Newton of force

$$\text{force} = \text{mass} \times \text{acceleration}$$
$$\therefore 1 \text{ newton} = 1 \text{ kg} \times 1 \text{ m/s}^2$$

It was stated in Chapter 1 that the acceleration due to gravity of a body falling freely was 9.81 m/s^2. Since the newton gives a 1 kilogram mass an acceleration of 1 metre per second squared, a kilogram mass falling freely will have acting upon it a force of 9.81 newton. This force is known as kilogram force and is used in calculations involving the pressure of water and other liquids.

Figure 2.2 shows a tank containing 1 cubic metre of water, which has a mass of approximately 1000 kg.

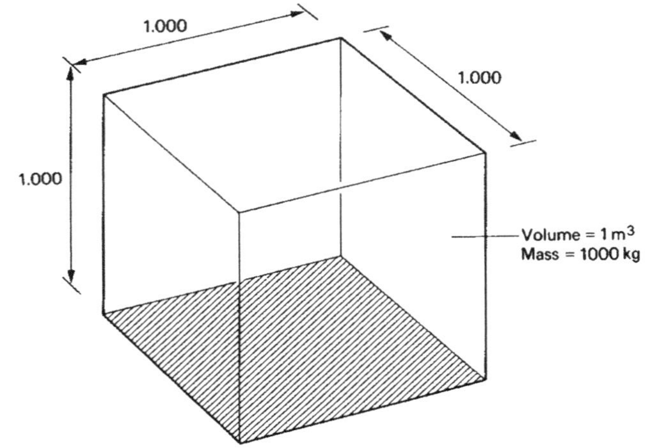

Pressure on the base due to 1 m head of water = 1000 x 9.81 N/m^2 = 9810 pascals or 9.81 kN/m^2 or 9.81 kPa

Fig. 2.2 Pressure due to 1 m^3 of water

The force acting over the base of 1 square metre due to the head of water of 1 metre will be:

force = mass × acceleration
= 1000 × 9.81
= 9810 N

Pressure of water

Since this force is acting over 1 square metre, the pressure on the base will be 9810 N/m^2, i.e.

$$\text{pressure} = \frac{N}{m^2} = N/m^2.$$

Since the pressure is proportional to head of water, the pressure for all calculations dealing with water will be:

N/m^2 = 1000 × 9.81 × head of water in metres

The newton per square metre is a small unit for water pressure and it is usually more convenient when dealing with water to use kilonewton per square metre.

kN/m^2 = 9.81 × head of water in metres

For liquids other than water the pressure may also be found because:

$$\text{pressure due to a liquid} = \text{head or height} \times \text{density} \times \text{acceleration due to gravity}$$

or

$$\text{pressure due to a liquid} = h\rho g$$

Example 2.1. *A rectangular tank measures 2 m long by 1.5 m wide by 2 m deep. The tank is filled with oil having a density of 900 kg/m^3 to a depth of 1.8 m. Find the pressure and the total force acting on its base.*

pressure = $h\rho g$
= 1.8 × 900 × 9.81
= 15 892.2 N/m^2
= 15.89 kN/m^2

force = pressure × area
= 15.89 × 2 × 1.5
= 47.67 kN

Note: $\frac{kN}{m^2} \times m^2 = kN$

The SI unit of pressure is known as the pascal, which is equal to 1 newton per metre squared, or

1 Pa = 1 N/m^2
1 kPa = 1 kN/m^2

Example 2.2. *Calculate the pressure of water acting upon a valve when the head of water above the valve is 15 m.*

pressure in Pa = head in metres × 1000 kg × 9.81
pressure in kPa = head in metres × 9.81
= 15 × 9.81
pressure = 147.15 kPa

When the pressure is known, it is sometimes necessary to find the head of water. Head is defined as the vertical distance between the free surface of the liquid and the point under consideration.

The pressure acting is proportional to head, and volume of liquid does not affect pressure. The pressure at the base of a tank holding 3 m^3 of water when the head is 1.5 m is exactly the same as the pressure on the base of the tank holding 150 m^3 with the same head of 1.5 m. The force acting on the base of the large tank, however, is much greater.

Example 2.3. *Calculate the pressure on the base of a tank measuring 4 m by 2 m by 2 m deep when the head of water is 1.5 m; also calculate the force acting on the base.*

pressure = 9.81 × 1.5
= 14.715 kPa or kN/m^3
$= \frac{\text{force}}{\text{area}}$

force = pressure × area
= 14.715 × 1.5
= 22.0725 kN
= 22 kN (approx.)

If the area is less than 1 metre squared, the pressure and force can still be found, as illustrated in example 2.4.

Example 2.4. *Calculate the pressure acting on a valve and the force acting on its horizontal seating of 50 mm radius when the head of water above the valve is 15 m.*

pressure acting on valve $= 9.81 \times 15$
$= 147.15$ kPa

force acting on valve $= 147.15 \times \pi R^2$
$= 147.15 \times 3.142 \times 0.05 \times 0.05$
$= 1.156$ kN

Example 2.5. *A pressure gauge on a boiler fed from a cistern shows a reading of 350 kPa. Calculate the head of water above the gauge.*

pressure = head in metres × 9.81

$$\text{head} = \frac{\text{pressure}}{9.81} = \frac{350}{9.81} = 35.68 \text{ m}$$

Absolute pressure

When dealing with problems in thermodynamics, absolute pressure is used.

$$\text{absolute pressure} = \text{gauge pressure} + \text{atmospheric pressure}$$

Atmospheric pressure at sea-level will support a column of mercury 760 mm high (see Fig. 2.3). Since mercury is approximately 13.6 times heavier than water, atmospheric pressure will support a column of water:

$760 \times 13.6 = 10\,336$ mm high
$= 10.33$ m
atmospheric pressure $= 10.33 \times 9.81 = 101.33$ kN/m^2
$= 101$ kPa (approx.)

Example 2.6. *Calculate the absolute pressure acting on top of a boiler when the head of water above the top is 20 m.*

gauge pressure $= 9.81 \times 20$
$= 196.2$ kPa
absolute pressure $= 196.2 + 101$
$= 297.2$ kPa

For ordinary pressure calculations, gauge pressure is used because atmospheric pressure is acting in opposite directions equally at the point under consideration and is therefore cancelled out. In example 2.6 the atmospheric pressure acting

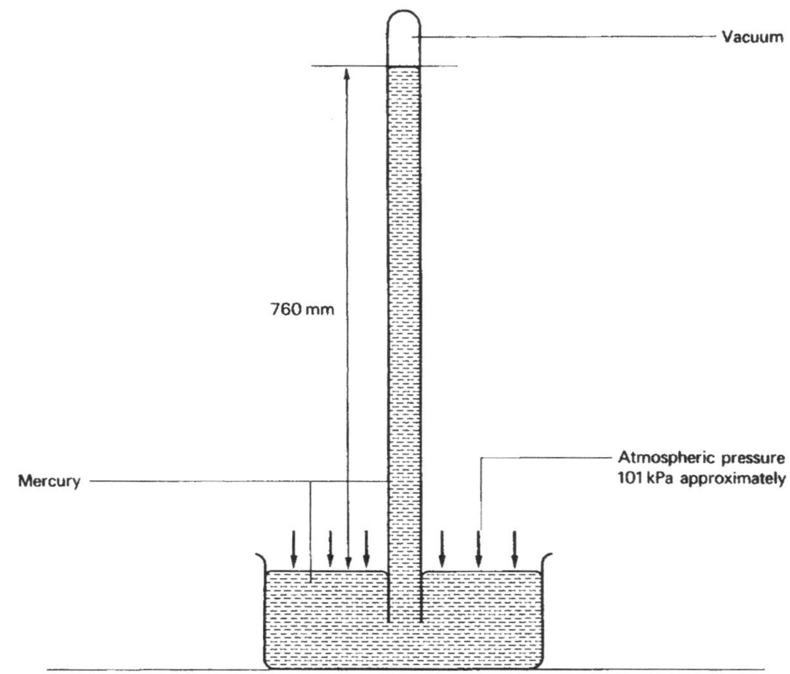

Fig.2.3 The mercury barometer

upon the surface of the water in the feed cistern is cancelled out by the atmospheric pressure acting outside on top of the boiler (see Fig. 2.4).

Example 2.7. *A water-filled manometer is connected to a gas pipe in order to measure the pressure of gas in the pipe. If the reading shows 200 mm head of water (water gauge), calculate the gas pressure in pascals and kilopascals.*

200 mm = 0.2 m head
pressure in pascals = head in metres × 1000 × 9.81
$= 0.2 \times 1000 \times 9.81$
$= 1962$ Pa
$= 1.962$ kPa

Gas pressure is sometimes measured in millibars or bars; these are not SI units but their values will be shown.

1 bar $= 10^5$ N/m^2 or 100 kN/m^2
1 mbar $= 100$ N/m^2 or 0.1 kN/m^2

$$\text{pressure in mbars} = \frac{1962}{100} = 19.62$$

$$\text{pressure in bars} = \frac{1.962}{100}$$

$$= 0.019\,62 \text{ bars}$$

Note: 1 bar is approximately equal to atmospheric pressure.

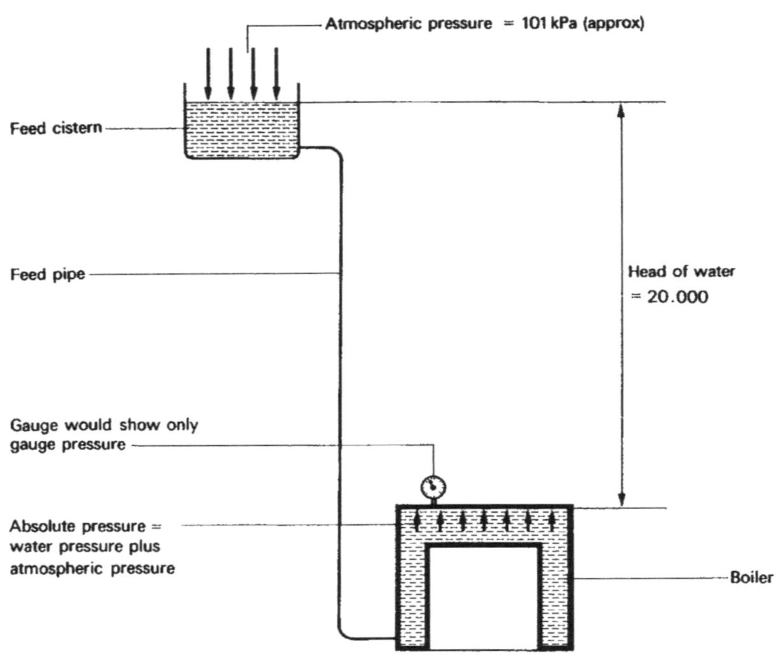

Fig.2.4 Absolute pressure on top of boiler

The hydraulic press or jack (see Fig. 2.5)

The hydraulic press is a machine which consists of two different size pistons, enabling it to lift weights by the application of a much smaller force. The weight that can be lifted depends upon the ratio of the areas of the two pistons. When the pistons are at the same height, the pressure of water acting upon both pistons is the same. The force acting on each piston is equal to the pressure multiplied by the area.

Example 2.8. *A hydraulic press has pistons having areas of 1000 mm² and 10 000 mm² respectively. What force is acting on the larger piston if a force of 10 kN is applied on the smaller one?*

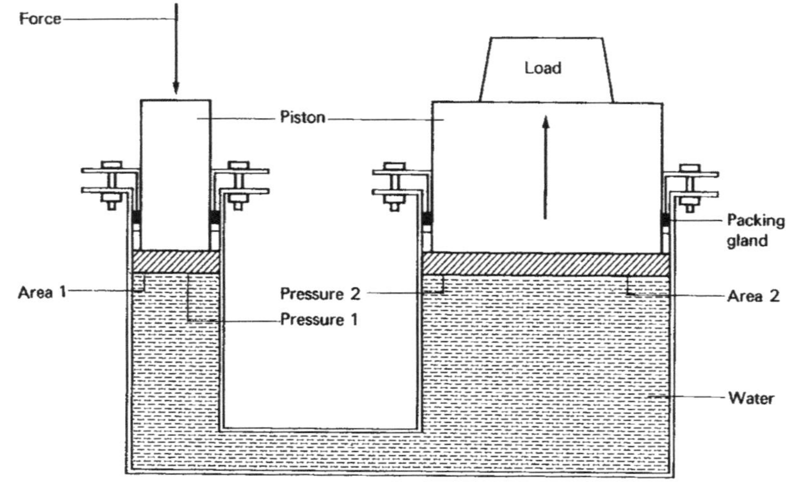

Fig.2.5 The hydraulic press

$$\text{ratio of pistons' areas} = \frac{10\,000}{1000}$$

$$= 1 : 10$$

Since water is virtually incompressible, the force acting upon the larger piston, neglecting friction, will be ten times the force acting on the smaller piston.

force acting to lift the load = 10 kN × 10

force on larger piston = 100 kN

Alternatively, the problem may be solved by finding the pressure acting upon the smaller piston.

$$\text{pressure} = \frac{\text{force}}{\text{area}} = \frac{\text{N}}{\text{mm}^2}$$

$$= \frac{10\,000}{1000}$$

$$= 10 \text{ N/mm}^2$$

Since this pressure is the same on the larger piston, the lifting force may be found.

$$\text{force} = \text{pressure} \times \text{area of larger piston}$$

$$= \text{N/mm}^2 \times \text{mm}^2$$

$$= 10 \times 10\,000 \text{ N}$$

$$\text{force on larger piston} = 100 \text{ kN}$$

Oil-hydraulic lift

The oil-hydraulic lift uses the principle of the hydraulic press. Oil at high pressure is pumped through a pipe to a cylinder containing a piston. The force acting upon the piston is equal to the pressure multiplied by the area of the piston. This force lifts the car plus its load.

Force on vertical surfaces (centre of pressure)

So far, formulae has been derived relating the head in metres with pressure in kilopascals and force on horizontal surfaces. Since water presses equally in all directions, consideration can now be given to surfaces which are vertical, such as the sides of cylinders, tanks, cisterns, access covers and sluice plates. The total force on any surface immersed in water is equal to the area of the surface multiplied by the pressure acting at the centre of the area (or the centroid) of the surface.

Example 2.9. (see Fig. 2.6): *A vertical rectangular plate is 1 m high by 1.200 m wide is acting as a sluice. The centre of the plate is 3 m below the surface of the water. Find the total force in kN acting on the plate.*

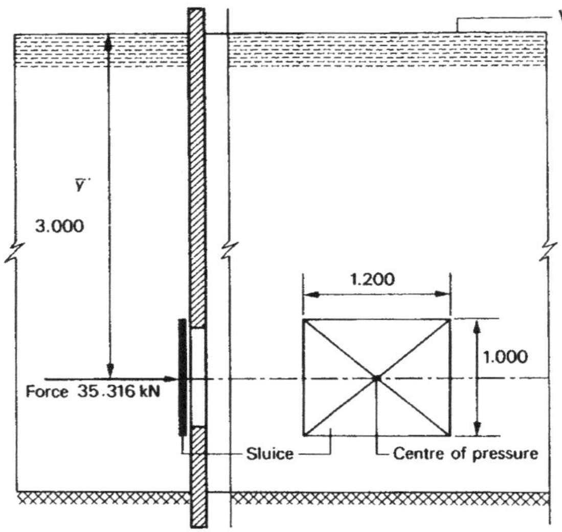

Note: The single resultant force acts through the centre of pressure

Fig.2.6 Force acting on a submersed plate

depth or head to centre of area $= \bar{y} = 3$ m

area of sluice plate $= 1 \times 1.2 = 1.200$ m^2

pressure at centre of plate $= \bar{y} \times 9.81$

$= 3 \times 9.81$

$= 29.43$ kPa

$$\text{pressure kPa} = \frac{\text{force kN}}{\text{area m}^2}$$

$\therefore$ force = pressure kPa $\times$ area m^2

$= 29.43 \times 1.200$

$= 35.316$ kN

Liquid contained in a vessel will exert a force at right angles to the surface of the vessel. The force will act upon each small element of area and the total force will be equal to the sum of these forces acting upon each small area.

total force acting = pressure at the centroid $\times$ total area.

Example 2.10 (see Fig. 2.7). *A water test is applied to a 100 mm diameter drain. Calculate the force acting upon the stopper when the head of water above the centre of the stopper is 4.05 m.*

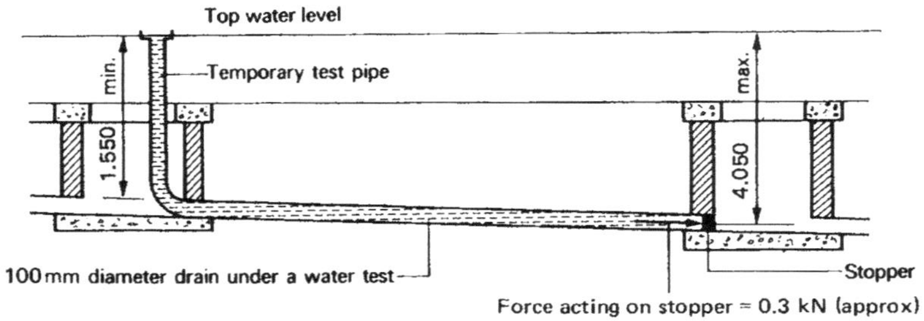

Fig.2.7 Force acting during drain testing

pressure at centroid $= \bar{y}$ (m) $\times 9.81$

$= 4.05 \times 9.81$

$= 39.73$ kPa

force acting upon stopper = pressure (kPa) $\times$ area (m^2)

$= 39.73 \times 3.142 \times 0.05 \times 0.05$

$= 0.312$ kN

$= 0.3$ kN (approx.)

Example 2.11 (see Fig. 2.8). *Assuming a static head of water, calculate: (a) the static pressure in kPa at A, B and C; (b) the force or thrust acting upon the circular manlid.*

pressure at A = head (m) $\times 9.81$

$= 4 \times 9.81$

$= 39.24$ kPa

pressure at B = head (m) $\times 9.81$

$= 7 \times 9.81$

$= 68.67$ kPa

pressure at C $= \bar{y}$ (m) $\times 9.81$

$= 10 \times 9.81$

$= 98.1$ kPa

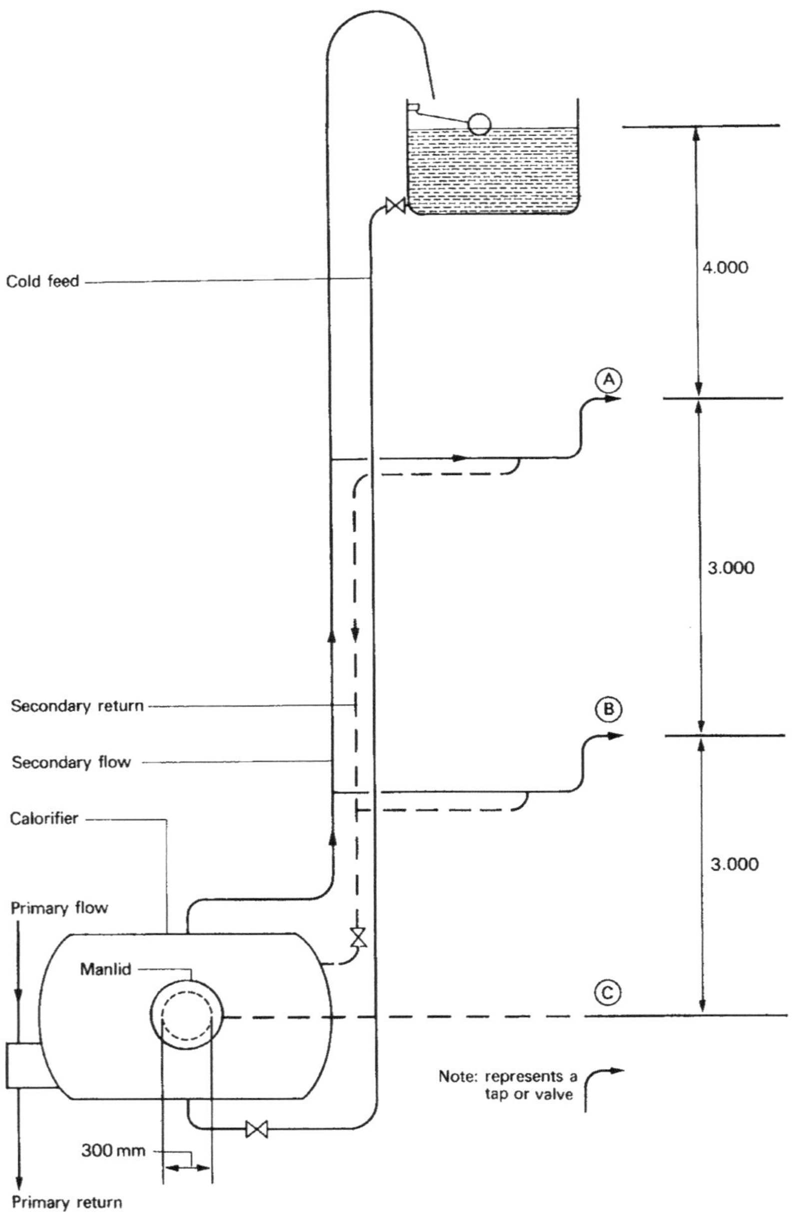

Fig. 2.8 Pressures in a hot-water system

force acting upon circular manlid = pressure (kPa) × area (m^2)
= 98.1 × 3.142 × 0.15 × 0.15
= 6.935 kN

Example 2.12 (see Fig. 2.8). *Calculate the force acting upon the total horizontal surface of the hot-water calorifier if its length is 2 m and internal diameter is 1.5 m.*

internal horizontal surface = πD × length
= 3.142 × 1.5 × 2
= 9.426 m^2

pressure at centre of calorifier = $\bar{y}$ (m) × 9.81
= 10 × 9.81
= 98.1 kPa

force = pressure × area
= 98.1 × 9.426
= 924.6906 kN
= 924.7 kN (approx.) (924 700 N)

This is a tremendous force and a cylindrical vessel will provide the greatest strength to resist it. A rectangular vessel would have to be constructed with a much greater thickness of metal.

Example 2.13 (see Fig. 2.9). *A rectangular cistern measures 3 m long by 2 m wide by 1.5 m deep and contains water to the depth of 1.4 m. Find by means of a pressure diagram the resultant force acting on one vertical end.*

Pressure increases proportionately with the depth. The pressure as shown in Fig. 2.9 is represented by a triangle increasing from zero at the top to maximum at the base.

pressure at the base = 1.4 × 9.81
= 13.734 kPa

Since the distribution of pressure is represented by a triangle, the resultant force acts through the centre of the area of the triangle, that is one-third of the height above the base.

The resultant force on the end may be found from the area of the pressure diagram.

area of triangle = $\frac{1}{2}$ AB × AC

resultant force per (m) run = $\frac{1}{2}$ × 1.4 × 13.734 kPa

force on end = $\frac{1}{2}$ × 1.4 × 13.734 × 2

resultant force = 19.2276 kN

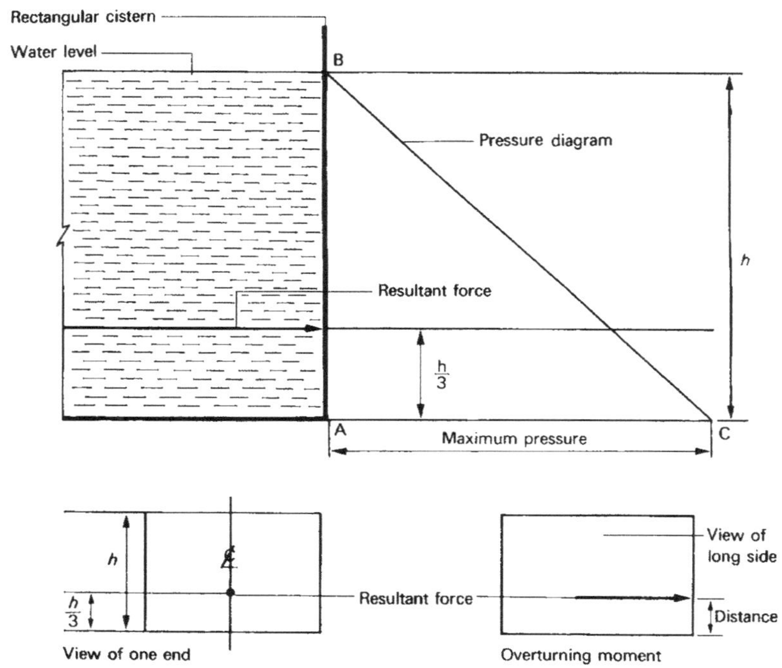

Fig.2.9 Pressure diagram showing the variation of pressure with depth

Check: The resultant force may also be found as follows:

$$\text{resultant force} = \text{pressure at } \frac{1}{2} \text{ depth} \times \text{area of end in contact with water}$$

$$= 0.7 \times 9.81 \times 1.4 \times 2$$
$$= 19.2276 \text{ kN}$$
$$= 19.23 \text{ kN (approx.)}$$

Overturning moment (see Fig. 2.9)

If the resultant force acting on the end of the cistern and the position where this force acts are known, the overturning moment which tends to overturn the end may be found.

$$\text{overturning moment} = \text{resultant force} \times \text{distance from base}$$

$$= 19.2276 \times \frac{1.4}{3}$$

$$= 8.972\,88 \text{ kNm}$$
$$= 9 \text{ kNm (approx.)}$$

In order to overcome this overturning moment, struts may be fixed across the top of the cistern. In order to resist the resultant force acting one-third of the head of water above the base, the metal should be made of sufficient strength and thickness, otherwise struts will also be required at this position.

Pressures on ballvalves

Ballvalves are classified according to the pressure that they are capable of closing against.

1. High-pressure valves should be capable of closing against a pressure of 1379 kPa.
2. Medium-pressure valves should be capable of closing against a pressure of 689.5 kPa.
3. Low-pressure valves should be capable of closing against a pressure of 276 kPa.

Every valve must be capable of withstanding a pressure of 2068.5 kPa when held in a closed position.

Questions

1. State the factors that decide the pressure exerted by a liquid.
2. Calculate the pressure in kPa acting on a valve when the head of water above the valve is 15 m.

Answer: 147.15 kPa

3. Calculate the force acting upon the base of a tank measuring 3 m by 2 m by 2 m deep. The tank contains water to a depth of 1.6 m.

Answer: 94.176 kN

4. Calculate the force acting upon a 76 mm diameter gate valve and the force acting upon the gate when the head of water above the valve is 20 m.

Answers: 196.2 kPa; 0.89 kN or 890 N

5. Calculate the force acting upon the top of a vertical calorifier when its diameter is 1.5 m and the head of water above the top is 5 m.

Answer: 86.678 kN or 86 678 N

6. Define the term 'centre of pressure'.
7. Describe how the total force acting upon one vertical side of a tank is calculated.
8. State why the metal struts placed across a tank full of water are placed near the top or one-third of the height from the base.
9. A pressure gauge on the top of a boiler gives a reading of 150 kPa. Calculate the head of water above the gauge.

Answer: 15.29 m

10. Define the terms 'absolute pressure' and 'gauge pressure'.

11. Calculate the 'absolute pressure' acting on the base of a tank when the head of water above the base is 12 m.

Answer: 117.72 kPa + 101 kPa = 218.72 kPa

12. A water-filled manometer is connected to a ventilating duct to measure the air pressure inside the duct. If the reading shows a head of water of 60 mm, calculate the air pressure in pascals.

Answer: 588.6 Pa

13. A water-filled manometer is connected to a gas pipe to measure the pressure of gas in the pipe. If the reading shows 150 mm head of water calculate the gas pressure in millibars.

Answer: 14.715 millibars

14. A rectangular tank measures 4 m long by 3 m wide by 2 m deep and contains water to a depth of 1.5 m.
    (*a*) Find, by means of a pressure diagram, the resultant force acting on one vertical end.
    (*b*) Show where the resultant force acts.

15. A pumping station is below ground level and the ground water rises to a maximum height, above the underside of the floor, of 1.5 m. If the floor has an area of 6 m^2, calculate the total force of water acting on the underside of the floor.

Answer: 88.29 kN

16. Water in a main has a pressure of 500 kPa and is to be used to supply a cold-water storage cistern having a ballvalve inlet 30 m above the centre of the main. Calculate the pressure acting on the ballvalve.

Answer: 205.7 kPa

17. A range of W.C. flushing cisterns are to be supplied from a cold-water storage cistern. The height from the water surface in the cistern and the ballvalves is 80 m. What type of ballvalve should be specified according to pressure classification?

Answer: high pressure (784.8 kPa)

18. A hydraulic press has pistons having areas of 500 mm^2 and 1500 mm^2 respectively. What load can be lifted on the larger piston if a force of 15 kN is applied to the smaller one?

Answer: 45 kN lifting force; 4.586 tonnes load lifted

Chapter 3
Hydraulics

Hydraulics

The subject of hydraulics concerns the study of fluids in motion. In order to maintain a certain rate of flow of fluid through a pipe or channel, a force is required to overcome the friction that exists between the surface of the pipe or channel and the moving fluid. In pipes this resistance is proportional to:

1. The square of the velocity (V^2).
2. The area of the surface in contact with the fluid.
   In the case of a pipe flowing full, this is equal to πdL, the curved surface of a cylinder,

   where d = diameter of pipe
   L = length of pipe
3. The nature of the internal surface.
   A rough surface will offer more resistance to flow than a smooth surface. Other factors that have an effect on flow are the density and viscosity of the fluid. If all the factors are constant, it has been shown by experiment that the discharge through a pipe is directly proportional to the square root of the fifth power of the diameter ($\sqrt{d^5}$).
   This is one of the discharge laws used in the Box formula, which will be described later in this chapter.

Relative discharging power of pipes

The relative discharging power of pipes are as the square root of the fifth power of their diameters.

$$N = \sqrt{\left(\frac{D}{d}\right)^5}$$

where N = number of branch pipes
D = diameter of main pipe
d = diameter of branch pipes

Example 3.1. *Calculate the diameter of a water main to supply twenty 32 mm diameter short branch pipes.*

$$N = \sqrt{\left(\frac{D}{d}\right)^5}$$

$$N^2 = \left(\frac{D}{d}\right)^5$$

$$\sqrt[5]{N^2} = \frac{D}{d}$$

$$D = \sqrt[5]{N^2} \times d$$

$$= \sqrt[5]{20^2} \times 32$$

$$= 106 \text{ mm}$$

The nearest commercial pipe size is 100 mm but this is undersized, so a 125 mm diameter pipe might be chosen.

Example 3.2. *Calculate the diameter of a water main to supply thirty 19 mm diameter short branch pipes.*

$$D = \sqrt[5]{N^2} \times d$$

$$= \sqrt[5]{30^2} \times 19$$

$$= 74 \text{ mm}$$

The nearest commercial pipe size is 76 mm, which would be satisfactory.

Example 3.3. *Calculate the number of 19 mm diameter short branch pipes that may be supplied from a 50 mm diameter main pipe.*

$$N = \sqrt{\left(\frac{D}{d}\right)^5}$$

$$= \sqrt{\left(\frac{50}{19}\right)^5}$$

$$= 11.24$$

Eleven branch pipes could be supplied.

Relative capacities of pipes

Although it is not a problem of hydraulics, it is sometimes necessary to compare the amount of water contained in pipes of various diameters. The relative capacities of pipes are as the square of their diameters.

$$N = \frac{D^2}{d^2}$$

where N = number of smaller pipes
d = diameter of smaller pipes
D = diameter of main pipe

Example 3.4. *How many 25 mm diameter branch pipes will be required to hold the same volume of water as one 76 mm diameter pipe?*

$$N = \frac{D^2}{d^2}$$

$$= \frac{76^2}{25^2}$$

$$= 9.243$$

Nine 25 mm diameter branch pipes would be required.

The relative discharging power of pipes takes into account the greater frictional loss in the branch pipes compared with the main pipe. A comparison of relative discharging power of pipes and relative capacities may therefore be made.

Example 3.2 showed that a 74 mm diameter main pipe would be required to supply thirty 19 mm diameter branch pipes. A comparison of relative capacities of pipes may be made using this example.

Example 3.5. *Calculate the number of 19 mm diameter pipes required to hold the same volume of water as one 74 mm diameter pipe.*

$$N = \frac{D^2}{d^2}$$

$$= \frac{74^2}{19^2}$$

$$= 15.17$$

It will be seen that about fifteen 19 mm branch pipes have the same capacity as one 74 mm diameter main pipe, whereas thirty 19 mm branch pipes have the same discharging power as one 74 mm diameter main pipe.

Types of flow (see Fig. 3.1)

When a fluid such as water flows in a pipe or channel, the flow may be either streamline or turbulent.

Streamline or laminar flow occurs when the motion of each molecule of fluid is in a straight line in the same direction as the axis of the pipe or channel.

Turbulent flow occurs when the motion of each molecule of the fluid is in a disorderly or disturbed manner.

Osborne Reynolds in 1883 recorded a number of experiments to determine the laws of resistance in pipes and whether the flow was streamline or turbulent. By introducing a filament of dye into the flow of water along a glass pipe, he showed that at low velocities the filament appeared in a straight line, indicating streamline flow. At high velocities, the filament, after passing a little way along the pipe, suddenly mixed with the surrounding water, indicating that the flow had become turbulent. The velocity of flow where the motion changes from streamline to turbulent is termed 'critical velocity'.

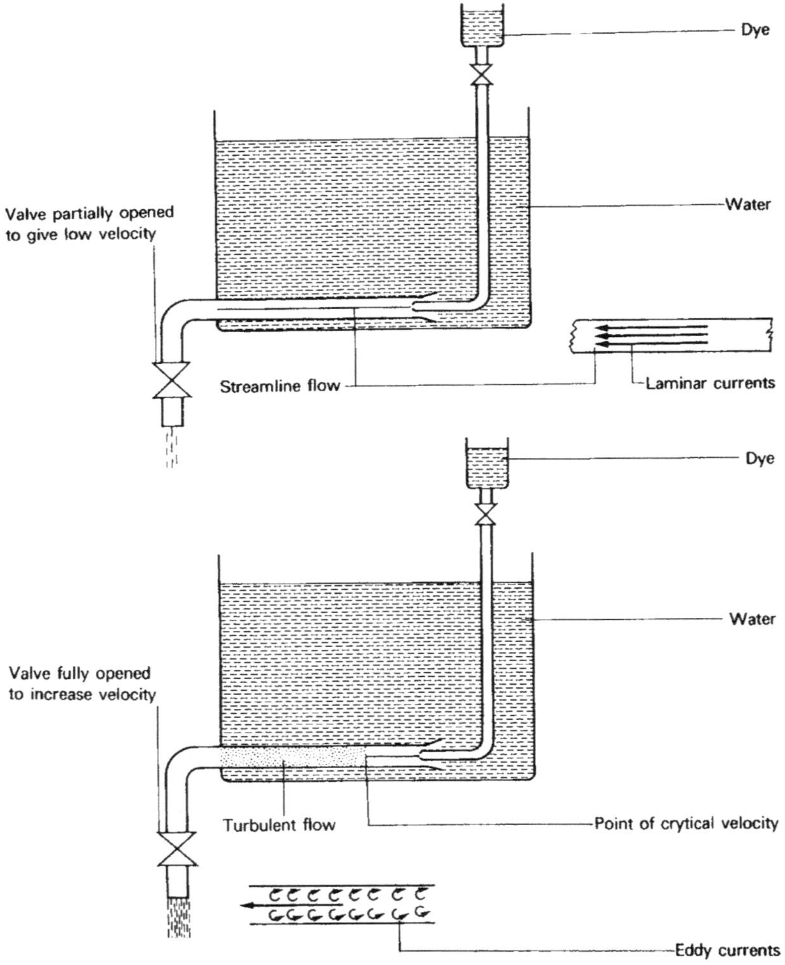

Fig.3.1 Streamline and turbulent flow

Reynolds number

Experiments with pipes of different diameters and with water at different temperatures led Reynolds to conclude that the type of flow was determined by:

(*a*) density of the fluid
(*b*) velocity of flow
(*c*) diameter of pipe
(*d*) viscosity of the fluid

Reynolds number $(R)\ \frac{\rho V d}{\nu}$

where ρ = density of fluid in kg/m^3
V = velocity in m/s
d = diameter of pipe in metres
ν = viscosity of the fluid in Pa s

Whatever the type of fluid or its temperature, if R is less than 2000 the flow will be streamline. If R is greater than 2800 the flow will be turbulent.

Example 3.6. *Calculate the Reynolds number for a fluid flowing through a 25 mm bore pipe at 2 m/s when its density is 900 kg/m^3 and its viscosity is 0.015 Pa s.*

$$R = \frac{\rho V d}{\nu}$$

$$= \frac{900 \times 2 \times 0.025}{0.015}$$

$$= 3000$$

This number would give turbulent flow which is commonly found in practical examples.

Streamline flow, however, prevents waste of energy and is therefore preferable in this respect. Higher velocities are used in order to reduce the diameters of pipes.

Pipe flow formula

The basic formula for the flow of water through pipes usually known as the D'arcy formula is given in the following expression:

$$h = \frac{4\,fLV^2}{2\,gD}$$

where h = loss of head due to friction in metres
f = coefficient of friction
L = length of pipe in metres
V = velocity of flow in metres per second
g = acceleration due to gravity in metres per second squared
D = diameter of pipe in metres

The value of the coefficient of friction depends upon the condition of the internal bore of the pipe. A value of 0.007 may be used for most problems and this is the value chosen by Thomas Box in his well-known formula which was derived from the D'arcy formula. The value may be found by experiment.

Example 3.7. *Find the loss of head due to friction in a 50 mm diameter pipe 30 m long when the velocity of flow is 2 m/s.*

$f = 0.007$
$g = 9.81$

$$h = \frac{4fLV^2}{2gD}$$

$$= \frac{4 \times 0.007 \times 30 \times 2^2}{2 \times 9.81 \times 0.05}$$

$$= 3.425 \text{ m}$$

Example 3.8. (see Fig. 3.2). *During an experiment conducted on a 25 mm bore copper pipe it was observed that the loss of head recorded by two water-filled manometers was 100 mm head of water. The tappings were 2 m apart and the velo~ity of flow 1 m/s. Calculate the coefficient of friction.*

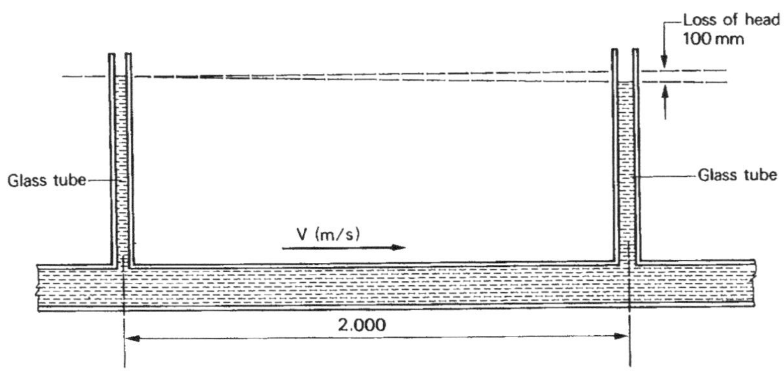

Fig.3.2 Loss of head due to friction

$$h = \frac{4fLV^2}{2gD}$$

by transposition

$$f = \frac{h \times 2g \times D}{4LV^2}$$

$$= \frac{0.1 \times 2 \times 9.81 \times 0.025}{4 \times 2 \times 1^2}$$

$$= 0.006\,131$$

Example 3.9. *Find the discharge in litres per second through a 32 mm bore pipe when the head lost due to friction is 5 m, length of pipe 20 m and the coefficient of friction is 0.007.*

$$h = \frac{4fLV^2}{2gD}$$

In order to find the discharge through the pipe, the velocity will be required.

By transposition

$$V = \sqrt{\frac{h\,2gD}{4fL}}$$

$$= \frac{5 \times 2 \times 9.81 \times 0.032}{4 \times 0.007 \times 20}$$

$$= \sqrt{\frac{3.139}{0.56}}$$

$$= 2.367 \text{ m/s}$$

$Q = VA$
where Q = flow rate in m^3/s
V = velocity of flow in m/s
A = area of pipe in m^2

$$Q = V \times 0.7854 \times D^2$$
$$= 2.367 \times 0.7854 \times 0.032 \times 0.032$$
$$= 0.0019 \text{ m}^3\text{/s}$$

discharge = 1.9 litre/s

Thomas box formula

This is a well-known formula for pipe sizing given in the following expression:

$$q = \sqrt{\frac{d^5 \times H}{25 \times L \times 10^5}}$$

where q = discharge through pipe in litres per second
d = diameter of pipe in millimetres
H = head of water in metres
L = total length of pipe in metres

As mentioned earlier, the formula was derived from the D'arcy formula in which the coefficient of friction 'f' chosen by Box was 0.007. This value is reasonably reliable for smooth pipes such as copper, plastic, stainless steel and lead. The coefficient of friction for galvanised mild steel pipes, however, is higher and therefore the Box formula is unreliable for these pipes.

Note: Steel pipes are graded as class A, B and C. Class A is a light grade and is not used for water services, but class B is used extensively. The internal diameter of class B (medium grade) is slightly greater than the internal diameter of a similar copper or plastic pipe. This is to permit threading of the pipe using a standard British pipe thread. This increase in diameter of class B pipe allows for a better flow of water than would otherwise be obtained and compensates for the greater resistance to flow of water. Class C pipe is heavy grade and is often used for underground water services.

Example 3.10 (see Fig. 3.3). *Two cold-water storage cisterns each containing 4500 litres are sited in a cistern room on the roof of a building and require to be refilled every two hours. The vertical height from the water main and the*

ballvalves is 26 m and the horizontal lengths from this main and the ballvalves is 9 m. If the pressure on the main supplying the cisterns is 300 kPa during the peak demand periods, calculate the diameter of the rising main assuming that the frictional losses on the pipe and fittings to be 30 per cent on the net length of pipe.

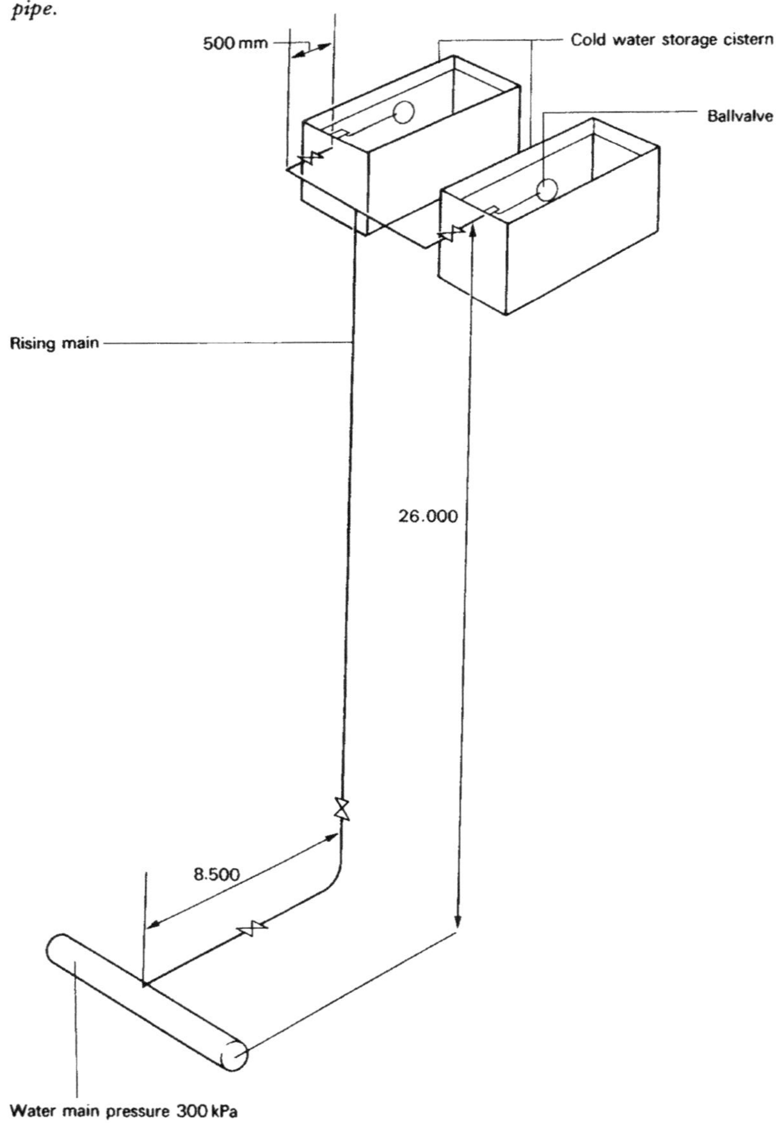

Fig.3.3 Size of rising main

$$\text{discharge in litres per second} = \frac{4500 \times 2}{2 \times 60 \times 60}$$
$$= 1.25$$

Head of water at ballvalves in metres (residual head)

pressure on water main = 300 kPa

$$\text{head of water on main} = \frac{300}{9.81}$$
$$= 30.58 \text{ m}$$

head lost due to height to ballvalves = 26 m

∴ residual head on ballvalves = 30.58 − 26 = 4.58 m

Length of pipe allowing 30 per cent for frictional losses

height = 26
horizontal length = 9
35

add 30 per cent
frictional resistance = 10.5
total length of pipe = 45.5 m

Values required in the Box formula:

1. discharge litre/s = 1.25
2. head of water in metres = 4.58
3. length of pipe in metres = 45.5

$$q = \sqrt{\frac{d^5 \times H}{25 \times L \times 10^5}}$$

$$q^2 = \frac{d^5 \times H}{25 \times L \times 10^5}$$

$$\text{by transposition } d^5 = \frac{q^2 \times 25 \times L \times 10^5}{H}$$

$$d = \sqrt[5]{\frac{q^2 \times 25 \times L \times 10^5}{H}}$$

$$\text{by substitution } d = \sqrt[5]{\frac{1.25^2 \times 25 \times 45.5 \times 10^5}{4.58}}$$

= 32.94 mm
= 33 mm (approx.)

The nearest commercial pipe size is 32 mm but this is slightly undersized; the next size is 38 mm which is 5 mm oversized; therefore a 32 mm diameter rising main would normally be chosen.

From an inspection of the Box formula it will be seen that the following pipe discharge laws will hold.

1. H varies directly with q^2.
2. H varies directly with L.
3. q varies directly with $\sqrt{H}$.
4. q varies directly with $\sqrt{d^5}$.

Variation of discharge

As the loss of head varies directly with the square of the discharge, then the following law of proportionality will hold:

$$H_1 \times q_1^2 = H_2 \times q_1^2$$

where H_1 = original head
H_2 = new head
q_1 = original discharge
q_2 = new discharge

Example 3.11. *If a 13 mm diameter pipe can discharge 0.15 litres per second with a head of water of 4 m, calculate the head of water required when discharging 0.3 litres per second.*

$$H_1 \times q_2^2 = H_2 \times q_1^2$$

$$H_2 = H_1 \times \left(\frac{q_2}{q^1}\right)^2$$

$$= 4 \times \left(\frac{0.3}{0.15}\right)^2$$

$$= 16 \text{ m}$$

Example 3.12. *If a 38 mm diameter pipe discharges 2 litres per second, calculate the discharge if the pipe run is increased to 76 mm diameter.*

$$q_1 \times \sqrt{d_2^5} = q_2 \times \sqrt{d_1^5}$$

$$q_2 = q_1 \times \sqrt{\left(\frac{d_2}{d_1}\right)^5}$$

$$= 2 \times \sqrt{\left(\frac{76}{38}\right)^5}$$

$$= 2 \times \sqrt{2^5}$$

$$= 11.314 \text{ litre/s}$$

$$= 11.3 \text{ litre/s (approx.)}$$

Example 3.13. *If a 13 mm diameter branch pipe is required to discharge 0.15 litres per second, calculate the size of main required to serve ten such branch pipes.*

total discharge = $10 \times 0.15 = 1.5$ litre/s

$$q_1 \times \sqrt{d_2^5} = q_2 \times \sqrt{d_1^5}$$

$$d_2 = \sqrt[5]{\left(\frac{q_2}{q_1}\right)^2} \times d_1$$

$$= \sqrt[5]{\left(\frac{1.5}{0.15}\right)^2} \times 13$$

$$= \sqrt[5]{10^2} \times 13$$

$$= 2.512 \times 13$$

$$= 32.656 \text{ mm}$$

$$= 33 \text{ mm (approx.)}$$

Example 3.14. *Calculate the discharge in litres per second through a 25 mm diameter pipe when the total length is 30 m and there is a constant head of 8 m.*

$$q = \sqrt{\frac{d^5 \times H}{25 \times L \times 10^5}}$$

$$= \sqrt{\frac{25^5 \times 8}{25 \times 30 \times 10^5}}$$

$$= 1.021 \text{ litre/s}$$

$$= 1 \text{ litre/s (approx.)}$$

Example 3.15. *Calculate the head of water necessary to provide a discharge of 0.2 litre/s through a 13 mm diameter pipe 6 m long.*

$$q = \sqrt{\frac{d^5 \times H}{25 \times L \times 10^5}}$$

by transposition $H = \dfrac{q^2 \times 25 \times L \times 10^5}{d^5}$

$$= \frac{0.2^2 \times 25 \times 6 \times 10^5}{13^5}$$

$$= 1.616 \text{ m}$$

$$= 1.6 \text{ m (approx.)}$$

Example 3.16. *Calculate the diameter of a pipe to discharge 1.25 litre/s when the constant head is 4 m and the total length of pipe 45.5 m.*

$$q = \sqrt{\frac{d^5 \times H}{25 \times L \times 10^5}}$$

by transposition $d = \sqrt[5]{\dfrac{q^2 \times 25 \times L \times 10^5}{H}}$

$$= \sqrt[5]{\frac{1.25^2 \times 45.5 \times 25 \times 10^5}{4}}$$

$$= 33.85 \text{ mm}$$

$$= 34 \text{ mm (approx.)}$$

Example 3.17. *Calculate the discharge in litres per second through a 32 mm diameter pipe when the head is 5 m and the total length 20 m.*

$$q = \sqrt{\frac{d^5 \times H}{25 \times L \times 10^5}}$$

$$= \sqrt{\frac{32^5 \times 5}{25 \times 20 \times 10^5}}$$

$$= 1.8 \text{ litre/s (approx.)}$$

Residual head (see Fig. 3.4)

The head of water calculated by the Box formula will provide the necessary pressure of water to overcome the resistance of water flowing through a pipe at a certain rate.

In practice, a pressure is also required to be left at the discharging point in order to operate certain fittings such as showers, sprinklers and flushing valves. The head of water left after overcoming friction in the pipe is known as the residual head and manufacturers will provide the required details for their fittings.

If in example 3.15 the residual head required is 1 m, then the total head would be:

head required to overcome friction + head required to operate the valve

= 1.616 + 1.000

= 2.616 m total head or effective head

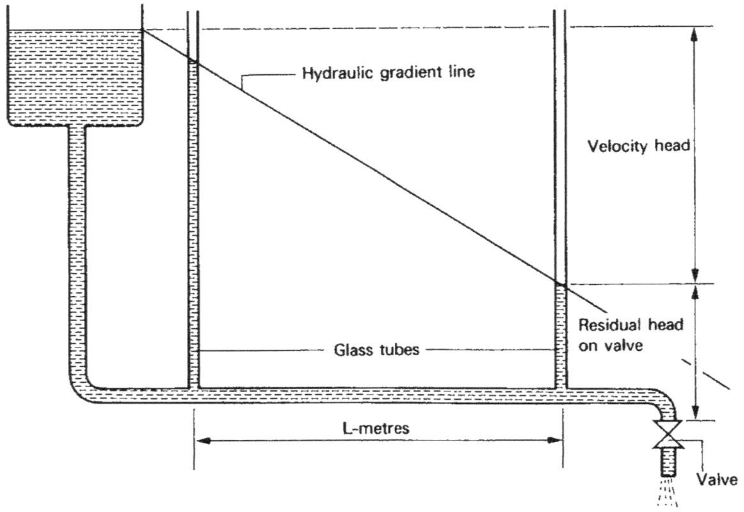

Fig.3.4 Residual head on valve

Resistance of fittings

The resistance to the flow of water due to fittings such as bends, elbows, tees, valves, etc. may be expressed in head of water.

The fundamental equation for velocity head given in Chapter 2 is

$$h = \frac{V^2}{2g}$$

Expressing the resistance of fittings in terms of velocity head the following may be used:

$$h = k\left[\frac{V^2}{2g}\right]$$

where h = velocity head lost due to fittings in metres
k = a number determined experimentally
V = velocity in m/s (velocity-head factor)
g = 9.81 m/s^2

Expressing a fitting in terms of velocity head,

$$k = \frac{h\,2g}{V^2}$$

k is not constant for any given fitting but varies with the velocity of flow. Providing a mean velocity is used when determining k, the variation above or below this velocity does not detract from the usefulness of this method of expressing resistance.

Resistance expressed as equivalent length of pipe

This method of allowing for resistance in terms of equivalent length of pipe is widely used by designers. These lengths are added to the net length of pipe to give the effective length used in either the D'arcy or the Box formula.

Table 3.1 gives values of k and equivalent lengths in pipe diameters for some common fittings and connections.

Table 3.1

Type of fitting or connection	Approx. value of k	Approx. equivalent length in pipe diameters
90° Elbow	1.00	30–36
Easy bend	0.3	10
Flush connection to cisterns, tanks, etc.	0.5	20
Globe valve	10	340
Tee		
straight	0.5	20
reducing one side	0.75	30
reducing two sides	1.00	36
water entering branch	2–2.5	70–90
Gate valve	0.2	7

Example 3.18. *Calculate the head lost due to friction through a globe valve when the velocity of flow is 2 m/s.*

$$h = k\frac{V^2}{2g}$$

$$= 10 \times \frac{2^2}{2 \times 9.81}$$

$$= 2.039 \text{ m}$$

Example 3.19. *Find the effective length of pipe when a 25 mm diameter pipe has the following fittings in its run, 2 globe valves, 4 bends and 2 straight tees. Net length of pipe 50 m.*

Resistance due to fittings:

globe valves	= 0.025 × 2 × 340 =	17 m
bends	= 0.025 × 4 × 10 =	1 m
tees	= 0.025 × 2 × 20 =	1 m
	total	19 m

effective length = 50 + 19 = 69 m

The Bernoulli theorem (see Fig. 3.5)

The theorem states that the total energy of each particle of a fluid is the same provided that no energy enters or leaves the system at any point. The theorem therefore is an aspect of the law of conservation of energy and if there is a loss of one type of energy there must be a corresponding gain in the other or vice versa.

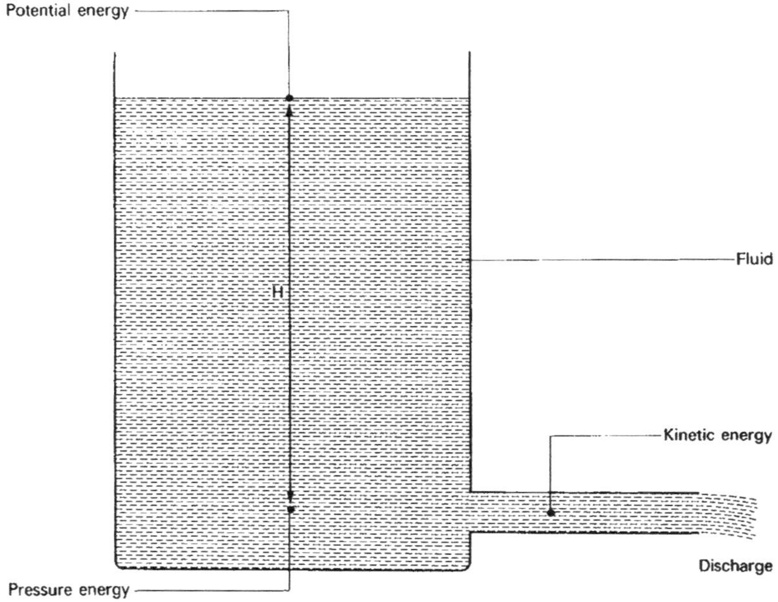

Fig. 3.5 The Bernoulli theorem

total energy $H = z + \frac{p}{\rho g} + \frac{V^2}{2g}$ = constant

where z = potential energy due to the position of the particles (m)

$\frac{p}{\rho g}$ = pressure energy due to the depth of the particles below the fluid (m)

$\frac{V^2}{2g}$ = kinetic energy due to the movement of the particles of the fluid (m)

In a closed system the total energy at only two points must be equal providing there are no gains or losses.

$$z_1 + \frac{p_1}{\rho g} + \frac{V_1^2}{2g} = z_2 + \frac{p_2}{\rho g} + \frac{V_2^2}{2g}$$

The Venturi meter (see Fig. 3.6)

A Venturi meter consists of a short inlet taper cone which reduces the pipe diameter to the throat diameter. This is followed by a divergent section which for minimum frictional loss should be longer than the inlet cone.

A 'U' tube is used to measure the pressure difference between the inlet and the throat.

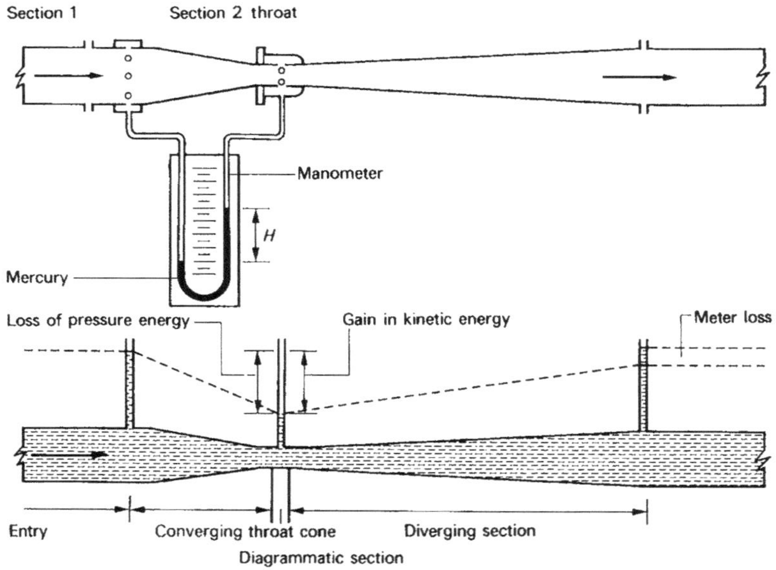

Fig. 3.6 The Venturi meter

Applying the Bernoulli equation:

$$z_1 + \frac{p_1}{\rho g} + \frac{V_1^2}{2g} = z_2 + \frac{p_2}{\rho g} + \frac{V_2^2}{2g}$$

For a horizontal meter $z_1 = z_2$ and are cancelled out of the two sides of the equation.

therefore $\frac{p_1}{\rho g} + \frac{V_1^2}{2g} = \frac{p_2}{\rho g} + \frac{V_2^2}{2g}$

and $\frac{V_2^2 - V_1^2}{2g} = \frac{p_1 - p_2}{\rho g}$

Since the same amount of fluid must pass through sections 1 and 2

$a_1 V_1 = a_2 V_2$

where a_1 = cross-sectional area of section 1
a_2 = cross-sectional area of section 2

By substitution and the addition of a coefficient of discharge, the expression for the flow in a Venturi meter can be shown:

$$Q = Cd\, a_1 \sqrt{\frac{2gH}{(a_1/a_2)^2 - 1}}$$

where Q = discharge in m^3/s
Cd = coefficient of discharge
a_1 = area of entrance pipe in m^2
a_2 = area of throat in m^2
H = differential head of water between inlet and throat in metres

Example 3.20. *A 225 mm diameter water main contains a Venturi meter whose throat diameter is 76 mm. A 'U' tube manometer containing mercury shows a difference in head between the inlet and the throat of 200 mm. If the coefficient of discharge is 0.95, calculate the rate of flow through the meter in m^3/s.*

Since mercury is approximately 13.6 times heavier than water, the head of water equivalent to 200 mm will be:

head of water in mm = 13.6 × 200
= 2720
head of water in metres = 2.72

The areas of the inlet and throat may be found from $0.7854\, D^2$

$a_1 = 0.7854 \times 0.225^2$
$= 0.0397\ m^2$
$a_2 = 0.7854 \times 0.076^2$
$= 0.0045$
$H = 2.72$
$g = 9.81$
$Cd = 0.95$

$$\text{discharge} = Cda_1 \sqrt{\frac{2gH}{(a_1/a_2)^2 - 1}}$$

$$= 0.95 \times 0.0397 \sqrt{\frac{2 \times 9.81 \times 2.72}{(0.0397/0.0045)^2 - 1}}$$

$$= 0.0377 \sqrt{\frac{53.366}{76.83}}$$

$= 0.0377 \times 0.833$
$= 0.0314\ m^3/s$ or 31.4 litre/s

Questions

1. Calculate by the relative discharging power formula the number of 13 mm diameter short branch pipes that may be supplied from a 25 mm diameter main pipe.

Answer: five 13 mm diameter pipes

2. Calculate the diameter of a main pipe to supply thirty 19 mm diameter short branch pipes.

Answer: 74 mm diameter

3. Describe streamline and turbulent flow. What is 'critical velocity' of flow?

4. Describe an experiment to find the coefficient of friction for the flow of water through a pipe.

5. Using the D'arcy formula, calculate the diameter of a pipe in millimetres when the length of pipe is 15 m, velocity of flow 3 m/s and the head lost due to friction 4 m. Use a coefficient of friction of 0.005 and $g = 9.81$.

Answer: 34.42 mm

6. If a 19 mm diameter pipe can discharge 4 litres per second with a head of water of 8 m, calculate the head of water when discharging 2 litres per second.

Answer: 2 m

7. If a 50 mm diameter pipe can discharge 8 litres per second, calculate the discharge if the pipe is reduced to 25 mm diameter.

Answer: 1.4 litre/s

8. Using the Box formula, calculate the diameter of a pipe in millimetres when the discharge required is 1.5 litres per second. The total length of pipe is 20 m and there is a constant head of water of 20 m.

Answer: 20.65 mm (a 25 mm diameter pipe would be used)

9. Using the Box formula, calculate the head of water required when the discharge is 5 litres per second through a 50 mm diameter pipe having a total length of 45 m.

Answer: 9 m

10. Calculate the diameter of a rising main using the following factors:
    (*a*) discharge required at the ballvalve 1.5 litres per second
    (*b*) pressure on the main during the peak demand period 300 kPa
    (*c*) height from the water main to the ballvalve 10 m
    (*d*) total length of pipe 20 m

Answer: 21 mm (a 25 mm diameter pipe would be suitable)

11. Define the term 'equivalent pipe length'.

12. Calculate the loss of head due to friction through fittings when a pipe run

contains 2 easy bends, 2 straight tees, and 1 gate valve. The velocity of flow is 2 metres per second.

Answer: 0.122 + 0.24 + 0.048 = 0.41 m

13. Define Reynolds number and state the factors required for its evaluation.
14. Define velocity head and residual head.
15. Explain the Bernoulli theorem and describe how the theorem may be used to measure the flow of water through a pipe.
16. A 300 mm diameter water main contains a Venturi meter whose throat diameter is 100 mm. A 'U' tube manometer containing mercury shows a difference in head between the inlet and the throat of 150 mm. If the coefficient of discharge is 0.98, calculate the rate of flow through the meter in m^3/s.

Answer: 0.0486 m^3/s or 48.6 litre/s

Chapter 4

Flow in drains and open channels, sewage disposal

Drains and open channels

Chezy formula

Various formulae, tables and charts may be used to find the gradient or velocity of flow of water in a drain or channel. One of the best-known formulae which may be used for pipes or channels is that due to Chezy, expressed as follows:

$$V = C\sqrt{m\,i}$$

where C = Chezy constant
V = velocity of flow (m/s)
m = hydraulic mean depth (m)
i = inclination or fall

The Chezy constant may be found from the following formula:

$$C = \sqrt{\frac{2g}{f}}$$

where g = acceleration due to gravity
f = coefficient of friction

For drainage, the average coefficient of friction may be taken as 0.0064 and therefore C would be given by

$$C = \sqrt{\frac{2 \times 9.81}{0.0064}}$$

$= 55$ (approx.)

$$m = \frac{\text{cross-sectional area of flow}}{\text{length of wetted perimeter}}$$

For drains flowing half- or full-bore, the hydraulic mean depth is equal to $D \div 4$, which can be shown as follows:

full bore $\quad m = \frac{\pi}{4} D^2 \div \pi D$

$$= \frac{\pi}{4} \times \frac{D^2}{1} \times \frac{1}{\pi} \times \frac{1}{D}$$

by cancellation $m = \frac{D}{4}$

half-bore $\quad m = \frac{\pi}{8} D^2 \div \frac{\pi D}{2}$

$$= \frac{\pi}{8} \times \frac{D^2}{1} \times \frac{2}{1} \times \frac{1}{\pi} \times \frac{1}{D}$$

by cancellation $m = \frac{D}{4}$

Figure 4.1 shows the relationship between the cross-sectional area of flow and the length of the wetted perimeter for half- and full-bore discharge.

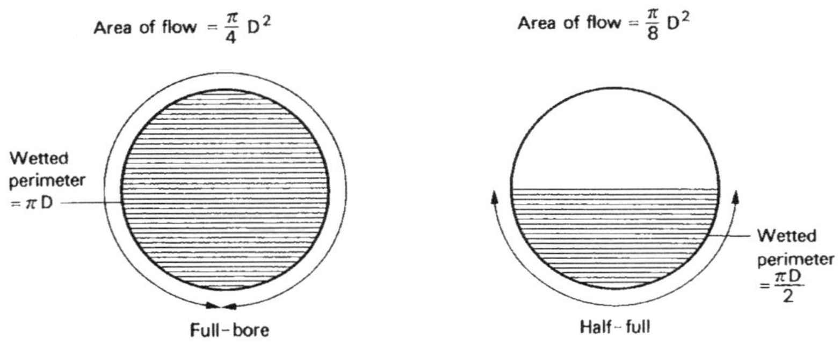

Fig.4.1 Hydraulic mean depth for full-bore or half-bore

The inclination, fall or gradient is equal to $H \div L$, where (L) is the length of drain in metres with a head (H) of 1 m (see Fig. 4.2).

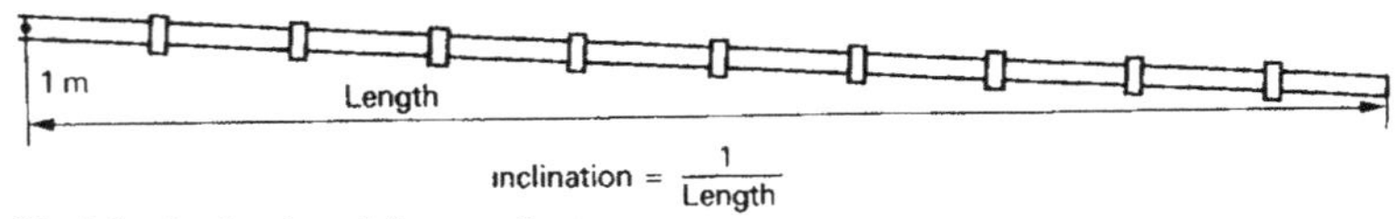

Fig.4.2 Inclination, fall or gradient

The Chezy formula was derived from the D'arcy formula described in Chapter 3. The coefficient of friction of 0.0064 is comparable to that used by Box in his pipe-sizing formula.

Example 4.1. *Using the Chezy formula, calculate the discharging capacity of a 150 mm diameter drain flowing full when laid to a fall of 1 in 140. (Take C = 55.)*

$$V = C\sqrt{m\,i}$$
$$= C\sqrt{\frac{D}{4} \times \frac{1}{L}}$$
$$= 55\sqrt{\frac{0.15}{4} \times \frac{1}{140}}$$
$$= 55\sqrt{0.016\,36}$$
$$= 0.9 \text{ m/s}$$

$Q = V A$

where Q = discharge in m^3/s
V = velocity of flow in m/s
A = area of pipe

$$\therefore \quad Q = V \times \frac{\pi}{4} D^2$$
$$= 0.9 \times \frac{3.142}{4} \times 0.15 \times 0.15$$
$$= 0.0159 \text{ m}^3\text{/s or } 15.9 \text{ litre/s}$$

Crimp and Bruges formula

This is another well-known formula often used by engineers, it is expressed as:

$$V = 84\, m^{2/3}\, i^{1/2}$$

or $\quad V = 84\sqrt[3]{m^2}\sqrt{i}$

where V = velocity of flow in m/s
m = hydraulic mean depth
i = inclination or fall

The formula may be used to calculate the velocity of flow in the 150 mm diameter drain laid at a gradient of 1 in 140 given in example 4.1.

$$V = 84\sqrt[3]{m^2}\sqrt{i}$$
$$= 84\sqrt[3]{\left[\frac{D}{4}\right]^2}\sqrt{\frac{1}{L}}$$
$$= 84\sqrt[3]{\left[\frac{0.15}{4}\right]^2}\sqrt{\frac{1}{140}}$$
$$= 84\sqrt[3]{0.0014} \times 0.0845$$
$$= 84 \times 0.1119 \times 0.0845$$
$$= 0.794 \text{ m/s}$$
$$= 0.8 \text{ m/s (approx.)}$$

Example 4.2. *Calculate the gradient required for a 100 mm diameter drain to run half-full at an average velocity of flow of 1.2 m/s. (Take C = 55.)*

$$V = C\sqrt{m\,i}$$

and $V = C\sqrt{\frac{D}{4} \times \frac{1}{L}}$

by transposition

$$\left[\frac{V}{C}\right]^2 = \frac{D}{4} \times \frac{1}{L}$$

$$\frac{1}{L} = \left[\frac{V}{C}\right]^2 \times \frac{4}{D}$$

$$L = \left[\frac{C}{V}\right]^2 \times \frac{D}{4}$$

by substituting the values given in the question

$$L = \left[\frac{55}{1.2}\right]^2 \times \frac{0.1}{4}$$

$$= 2100.69 \times 0.025$$
$$= 52.52 \text{ (approx.)}$$

A gradient of 1 in 52.5 would be required.

Note: For flows in channels having other than circular cross-sections, the hydraulic mean depth may also be calculated.

Example 4.3. *Figure 4.3 shows a cross-section of a rectangular channel. Calculate the gradient if the average velocity of flow is to be 0.8 m/s.*

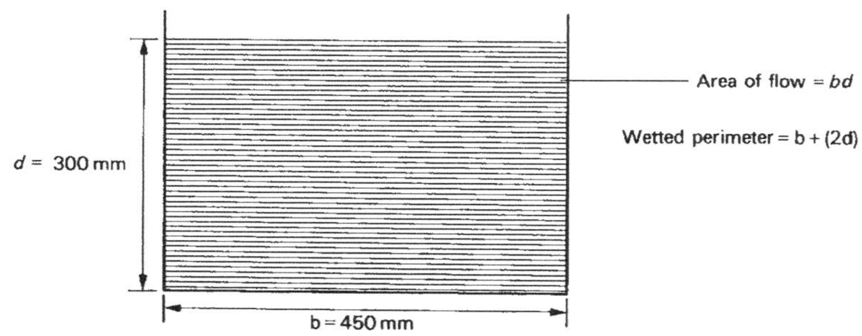

Fig.4.3 Hydraulic mean depth for a rectangular channel

the hydraulic mean depth $= \frac{b\,d}{b + (2d)}$

$$= \frac{0.45 \times 0.3}{0.45 + (2 \times 0.3)}$$

$$= \frac{0.135}{0.45 + 0.6}$$

$$= \frac{0.135}{1.05}$$

$$= 0.1286$$

$$V = C\sqrt{m\,i}$$

$$= C\sqrt{m \times \frac{1}{L}}$$

by transposition $\left[\frac{V}{C}\right]^2 = m \times \frac{1}{L}$

$$\left[\frac{V}{C}\right]^2 \times \frac{1}{m} = \frac{1}{L}$$

$$L = \left[\frac{C}{V}\right]^2 \times m$$

$$= \left[\frac{55}{0.8}\right]^2 \times 0.1286$$

$$= 4726.5625 \times 0.1286$$
$$= 607.83$$

A gradient of 1 : 608 (approx.) is required.

Value of (m) for surface and foul-water drains

It is usual practice to design surface-water drains to run full-bore and foul-water drains to run at a maximum depth equal to three-quarters of the internal diameter of the pipe.

It has been shown that the hydraulic mean depth for full-bore discharge is equal to the diameter divided by four (or multiplied by 0.25).

For foul-water drains it will be necessary to find the hydraulic mean depth before the velocity of flow or the gradient may be found by the Chezy formula.

Example 4.4 (see Fig. 4.4). *Calculate the velocity of flow in a 100 mm diameter foul-water drain flowing at a depth equal to three-quarters of the diameter when the gradient is 1 in 60; Chezy constant 55.*

area of water flow = area of segment of circle plus area of triangle

area of segment of circle $= \frac{\theta}{360} \times \frac{\pi}{4} D^2$

$$= \frac{240}{360} \times \frac{3.142}{4} \times \frac{0.1}{1} \times \frac{0.1}{1}$$

$$= 0.005\ 24 \text{ m}^2 \text{ (approx.)}$$

area of triangle $= \frac{b\,h}{2}$

$$= \frac{0.086 \times 0.025}{2}$$

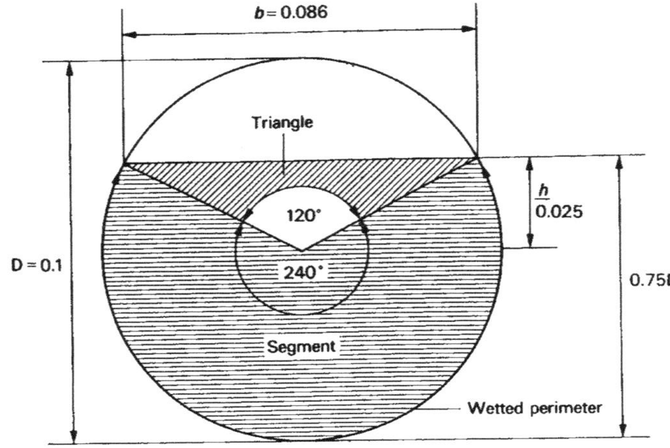

Fig.4.4 Hydraulic mean depth for $\frac{3}{4}$ depth

$$= 0.0011 \text{ m}^2 \text{ (approx.)}$$

$$\text{area of water flow} = 0.005\,24 + 0.0011$$

$$= 0.006\,34 \text{ m}^2$$

Wetted perimeter

$$\text{arc of circle} = \frac{\theta}{360} \times \pi D$$

$$= \frac{240}{360} \times 3.142 \times 0.1$$

$$= 0.2095 \text{ m (approx.)}$$

$$\text{hydraulic mean depth} = \frac{\text{area of water flow}}{\text{wetted perimeter}}$$

$$= \frac{0.006\,34}{0.2095}$$

$$= 0.0303 \text{ (approx.)}$$

Using Chezy formula to find velocity,

$$V = C\sqrt{m\,i}$$

$$= 55\sqrt{0.0303 \times \frac{1}{60}}$$

$$= 55 \times 0.0224$$

$$= 1.236 \text{ m/s (approx.)}$$

Table 4.1 may be used to find the hydraulic mean depth for pipes with water flowing at various depths.

Table 4.1

Depth of flow	Hydraulic mean depth
Full or $\frac{1}{2}$ full bore	diameter × 0.25
$\frac{3}{4}$ depth of flow	diameter × 0.30
$\frac{2}{3}$ depth of flow	diameter × 0.29
$\frac{1}{3}$ depth of flow	diameter × 0.19
$\frac{1}{4}$ depth of flow	diameter × 0.15

Note: The approximate maximum velocity of flow of water through a drain is obtained when the pipe is flowing at a depth equal to four-fifths of the internal diameter. The minimum self-cleansing velocity of flow is 0.8 m/s, but there is no upper limit for self-cleansing velocity.

Design of surface water drainage

Rainfall intensity. For storms expected to occur once a year the following formulae may be used.

(*a*) $R = \frac{750}{(T + 10)}$

where T is 5 to 10 minutes

(*b*) $R = \frac{1000}{(T + 20)}$

where T is 20 to 100 minutes

In these formulae, R = rainfall per hour in mm
T = duration of storm in minutes

Note: The maximum run-off at the point of outgo does not occur until the end of the time of concentration, when the whole area is contributing to the discharge. Where a time of concentration is used, such a time must be added to T in the formulae.

Rainfall run-off

The amount of water which can be expected from any given surface depends upon the following factors:

1. the area of the surface;
2. the type of surface;
3. whether the surface is level or sloping;
4. the intensity of rainfall;
5. the seasonal conditions, i.e. rate of evaporation.

The following formula may be used to find the volume of water run-off:

$$Q = \frac{A\,P\,R}{60 \times 1000 \times 60}$$

where Q = run-off in m^3/s
A = area drained in m^2

P = impermeability factor
R = rainfall intensity in mm/hr

Table 4.2 gives the impermeability factors for various types of surfaces.

Table 4.2

Type of surfaces	**Impermeability factor (average)**
Watertight surfaces	0.90
Asphalt pavements in good order	0.875
Closely jointed wood or stone pavements	0.825
Macadam roadways	0.425
Lawns and gardens	0.15
Wooded areas	0.105

Example 4.5. *Calculate the diameter and gradient of a pipe required to carry the run-off from an area of one hectare of lawns and gardens using the following factors:*

1. duration of storm 5 minutes;
2. impermeability factor 0.15;
3. time of concentration 18 minutes;
4. full-bore discharge;
5. velocity of flow 0.8 m/s;
6. Chezy constant 55.

Note: 1 hectare = 10 000 m^2. The time of concentration is not usually used unless it exceeds about 15 minutes.

$$R = \frac{750}{T + t + 10}$$

where R = rainfall intensity (mm/hr)
T = duration of storm in minutes
t = time of concentration in minutes

$$\therefore R = \frac{750}{5 + 18 + 10} = \frac{750}{33} = 22.73 \text{ mm/hr}$$

$$Q = \frac{A\,P\,R}{60 \times 1000 \times 60} = \frac{10\,000 \times 0.15 \times 22.73}{60 \times 1000 \times 60} = \frac{34\,095}{3\,600\,000} = 0.009\,47 \text{ m}^3/\text{s}$$

also $Q = V A$

$$= V \times \frac{\pi}{4} D^2$$

$$D = \sqrt{\frac{4\,Q}{V \pi}} = \sqrt{\frac{4 \times 0.009\,47}{0.8 \times 3.142}} = \sqrt{\frac{0.037\,88}{2.5136}} = 0.123$$

A 125 mm diameter pipe would be satisfactory.

To find the gradient:

$$L = \left[\frac{C}{V}\right]^2 \times \frac{D}{4} = \left[\frac{55}{0.8}\right]^2 \times \frac{0.125}{4} = 4726.56 \times 0.031\,25 = 147.705$$

A gradient of 1 in 148 would be satisfactory.

If flooding is liable to cause damage or inconvenience, a rainfall intensity of 50 mm/hr is often used.

Example 4.6 (see Fig. 4.5). *Calculate the diameter and gradient of a pipe required to carry the run-off from an asphalt-covered car park measuring 50 m × 30 m using the following factors:*

1. impermeability factor 0.875;
2. full-bore discharge;
3. rainfall intensity 50 mm/hr;
4. velocity of flow 1.5 m/s;
5. Chezy constant 55.

$$Q = \frac{A\,P\,R}{60 \times 1000 \times 60} = \frac{50 \times 30 \times 0.875 \times 50}{60 \times 1000 \times 60} = \frac{65\,625}{3\,600\,000} = 0.018\,229 \text{ m}^3/\text{s}$$

$$= 0.0182 \text{ m}^3/\text{s (approx.)}$$

Alternatively, Q may be found from the following formula:

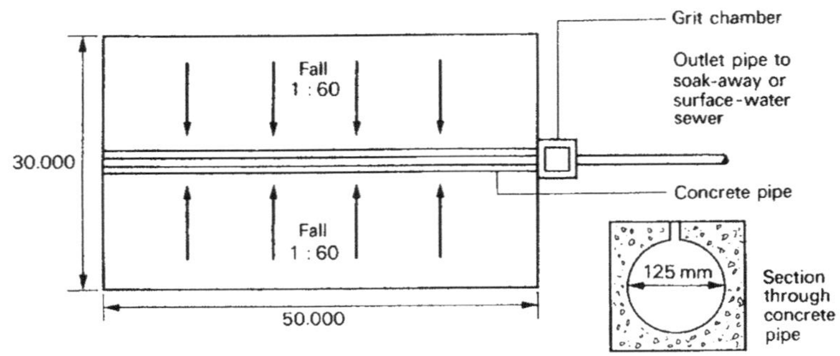

Fig.4.5 Plan of surface-water drainage for an asphalt covered car park

$$Q = \frac{\text{area to be drained (m}^2) \times \text{rainfall intensity (m/h)} \times \text{impermeability factor}}{3600}$$

$$= \frac{50 \times 30 \times 0.05 \times 0.875}{3600}$$

$$= 0.0182 \text{ m}^3\text{/s (approx.)}$$

$$Q = VA;\ Q = V\frac{\pi}{4}D^2;\ D = \sqrt{\frac{4Q}{V\pi}}$$

$$D = \sqrt{\frac{4 \times 0.0182}{1.5 \times 3.142}}$$

$$= 0.125 \text{ m}$$
$$= 125 \text{ mm}$$

To find the gradient:

$$L = \left[\frac{C}{V}\right]^2 \times \frac{D}{4}$$

$$= \left[\frac{55}{1.5}\right]^2 \times \frac{0.125}{4}$$

$$= 1344.444 \times 0.031\,25$$
$$= 42$$

gradient = 1 in 42

Size of soakaway

A soakaway may be sized on the basis of a rainfall intensity of 15 mm per hour. Assuming that three soakaways are to be used to take the discharge from the car park in example 4.6, the size of each may be calculated as follows:

$$\text{volume of each soakaway} = \frac{50 \times 30 \times 0.015}{3}$$

$$= 7.5 \text{ m}^3$$

assuming each soakaway is square on plan and 2 m deep,

$$\text{volume} = \text{area} \times \text{depth}$$

$$\text{area} = \frac{\text{volume}}{\text{depth}}$$

$$\text{length of side} = \sqrt{\frac{\text{volume}}{\text{depth}}}$$

$$= \sqrt{\frac{7.5}{2}}$$

$$= 1.94 \text{ m}$$

$\therefore$ size of each soakaway = 2 m × 2 m × 2 m

Alternatively, two soakaways may be used

$$\text{volume of each soakaway} = \frac{50 \times 30 \times 0.015}{2}$$

$$= 11.25 \text{ m}^2$$

Assuming each soakaway is square on plan and 2 m deep,

$$\text{length of one side} = \sqrt{\frac{11.25}{2}}$$

$$= 2.372 \text{ m}$$

$\therefore$ size of each soakaway = 2.372 m × 2.372 m × 2 m

Foul-water drains

The diameter of a foul-water drain will depend upon the discharge from the sanitary fittings during the peak demand period. The following factors will have to be considered:

1. the number and types of sanitary fittings;
2. the possible peak frequency of use of the fittings;
3. the average duration of discharge of the fittings;
4. the volume of water discharged from the fittings.

Discharge-unit method

When the number and types of sanitary fittings are known, the diameter and gradient of a foul-water drain may be found by the discharge-unit method. By this method, each appliance is given a discharge-unit value which represents the discharge capacity and frequency of use of the fitting. The sum of all the discharge units of the sanitary fittings are found.

Table 4.3 gives a list of discharge units and Table 4.4 the number of discharge units to be allowed on 100 mm, 125 mm and 150 mm diameter horizontal drains and their gradients.

Example 4.7. *By use of the discharge-unit method, find the diameter and gradient of a horizontal drain to serve a five-storey office containing 8 W.Cs,*

Table 4.3 Discharge-unit values BS 5572 1978

Type of sanitary fitting	Interval between uses			
	Domestic (min)	Public (min)	Peak flow (min)	Discharge unit value
Automatic washing machine				4
		10		20
			5	40
WC (9-litre cistern)	20			7
		10		14
			5	28
Sink	20			6
		10		14
			5	27
Wash basin	20			1
		10		3
			5	6
Bath	75			7
		30		18
			—	—
Shower (per head)				
domestic				1
public				2
Urinal (stall or bowl)		20		0.3

Table 4.4 Maximum number of discharge units to be allowed on horizontal branches BS 5572 1978

Internal diameter of pipe (mm)	Fall		
	1 in 111	1 in 45	1 in 22
100	230	430	1050
125	780	1500	3000
150	2000	3500	7500

8 basins, 2 sinks, and 4 urinals on each floor. The W.Cs are to be provided with 9 litre flushing cisterns.

W.Cs	8 × 5 × 14 DUs =	560
Basins	8 × 5 × 2 DUs =	120
Sinks	2 × 5 × 14 DUs =	140
Urinals	4 × 5 × 0.3 DUs =	6
	Total	826

With reference to Table 4.4, a 150 mm diameter drain having a gradient of 1 in 111 would be suitable.

Simultaneous demand factor

The discharge-unit value takes into account the number of fittings discharging simultaneously compared with the total number of fittings installed (usually expressed as a percentage). The number of fittings discharging simultaneously, during the peak demand period, may be found by the application of the theory of probability and a simplified formula may be used, given by the expression:

$$m = np + 1.8\,[2\,np\,(1 - p)]^{0.5}$$

where m = number of fittings discharging simultaneously

$p = \frac{t}{T}$ known as the usage ratio

n = total number of fittings installed
t = time in seconds fittings are discharging
T = time in seconds between usage

Example 4.8. *Calculate the number of fittings discharging simultaneously if 100 fittings, each of which take 10 seconds to discharge their contents, are used at 400-second intervals. Calculate the simultaneous demand factor.*

$$p = \frac{t}{T}$$

$$\therefore p = \frac{10}{400} = 0.025$$

$$m = (100 \times 0.025) + 1.8\,[2 \times 100 \times 0.025\,(1 - 0.025)]^{0.5}$$
$$= 4.3 + \sqrt{4.875}$$
$$= 4.3 + 2.2$$
$$= 6.5$$

simultaneous demand factor (per cent)

$$= \frac{\text{number of fittings discharging simultaneously}}{\text{number of fittings installed}} \times \frac{100}{1}$$

$$= \frac{6.5}{100} \times \frac{100}{1}$$

$$= 6.5 \text{ per cent}$$

If the simultaneous demand factor and the rate of discharge from the fittings are known, the diameter and the gradient of a foul-water drain may be calculated.

Table 4.5 gives the approximate rate of discharge from various sanitary fittings.

Example 4.9. *If the fittings given in example 4.8 consist of 40 W.Cs, 50 basins and 10 sinks with 38 mm diameter wastes, calculate the diameter and gradient of a drain to carry their discharges when the velocity of flow is to be 1 m/s and the drain is to flow at a maximum of half-full bore. Chezy constant = 55.*

Discharge from fittings (see Table 4.5):

Table 4.5 Rate of discharge from sanitary fittings

Type of fitting	Discharge (litres per second)
Basin	0.60
Bath	1.00
Sink (32 mm diameter waste)	0.50
Sink (38 mm diameter waste)	1.00
Urinal	0.15
WC (9 or 14 litre cistern)	2.30
Shower	1.00

W.Cs 40 × 2.3 = 92
Basins 50 × 0.6 = 30
Sinks 10 × 1.0 = 10
Total 132 litre/s

allowing 6.5 per cent simultaneous demand factor $= \frac{132}{1} \times \frac{6.5}{100}$

$$\text{discharge} = 8.58 \text{ litre/s} = 0.008\,58 \text{ m}^3\text{/s}$$

$$Q = V A$$

$$= V \times \frac{\pi}{8} D^2 \text{ (half-full bore)}$$

$$D = \sqrt{\frac{Q \times 8}{V \times \pi}}$$

$$= \sqrt{\frac{0.008\,58 \times 8}{1 \times 3.142}}$$

$$= 0.1478 \text{ m} = 148 \text{ mm}$$

A 150 mm diameter pipe would be satisfactory.

To find the gradient:

$$L = \left[\frac{C}{V}\right]^2 \times \frac{D}{4}$$

$$= \left[\frac{55}{1}\right]^2 \times \frac{0.15}{4}$$

$$= 3025 \times 0.0375 = 113.44$$

A gradient of 1 in 113 would be satisfactory.

If the discharge-unit method is used to find the diameter and gradient for the drain given in example 4.9, these would be as follows:

W.Cs 40 × 14 DUs = 560
Basins 50 × 3 DUs = 150
Sinks 10 × 14 DUs = 140
Total 850

With reference to Table 4.4, a 150 mm diameter drain having a gradient of 1 in 96 would be suitable.

Foul-water sewers

Experience has shown that, for the design of foul-water sewers on the separate system of drainage, a basis of six times the average 24-hour flow, determined by the water consumption per head of population, generally provides an adequate factor of safety.

The approximate water consumptions are: 140; 180; 240 litres per person per day.

Example 4.10. *Calculate the diameter of a foul-water sewer suitable for 150 housing units. The average number of occupants per house is assumed to be four and the water consumption 240 litres per person per day. The sewer is to be designed to run half-full bore, with a velocity of flow of 1 m/s.*

rate of flow per day = 240 × 150 × 4 = 144 000 litres

average flow rate during a six-hour period assumed half the daily flow = 72 000 litres

average flow rate per hour $= \frac{72\,000}{6} = 12\,000$ litres

average flow rate per second $= \frac{12\,000}{60 \times 60} = 3.333$

maximum flow rate per second assuming four times the average $= 3.333 \times 4 = 13.333$ litre/s $= 0.0133$ m^3/s

$$Q = V A = \frac{V \pi D^2}{8}$$

$$D = \sqrt{\frac{8\,Q}{V \pi}}$$

$$= \sqrt{\frac{8 \times 0.0133}{1 \times 3.142}}$$

$$= 0.184 \text{ m} = 184 \text{ mm}$$

A 200 mm diameter pipe would be satisfactory.

The approximate peak flow rates for three different water consumptions are 140, 180 and 240 litres per person per day. If 180 litres per person per day was chosen in example 4.10, the diameter of the sewer would be calculated in example 4.11.

Example 4.11.

rate of flow per day $= 180 \times 150 \times 4$
$= 108\ 000$ litres

average flow rate during six-hour period assuming half the daily flow $= 54\ 000$ litres

average flow rate per hour $= \dfrac{54\ 000}{6}$
$= 9000$ litres

average flow rate per second $= \dfrac{9000}{60 \times 60}$
$= 2.5$

maximum flow rate per second assuming four times the average $= 2.5 \times 4$
$= 10$ litre/s
$= 0.01\ m^3/s$

$$D = \frac{8\,Q}{V\,\pi}$$

$$= \sqrt{\frac{8 \times 0.01}{1.2 \times 3.142}}$$

$= 0.146$ m
$= 146$ mm

A 150 diameter pipe would be satisfactory.

The **CP 301**, Building Drainage, considers that not more than, say, 20 housing units should be connected to a 100 mm nominal bore drain and not more than 150 housing units should be connected to a 150 mm diameter nominal bore drain. Example 4.11 shows that a 150 mm nominal bore pipe is suitable when the water consumption is taken as 180 litres per person per day and a velocity of flow is 1.2 m/s.

Drainage above ground

Terminal velocity in stacks

It was shown in Chapter 1 that if a body is falling freely near to the surface of the earth it has an acceleration due to gravity of 9.81 m/s^2. When water flows down a vertical drainage stack, however, the forces of gravity and friction acting on the flow, soon balance and a terminal velocity is reached (see Fig. 4.6).

The height of stack necessary for terminal velocity to be achieved depends upon the diameter of the pipe, amount of flow and smoothness of the internal bore. The height is only likely to be equal to one storey. There is, however, no precaution necessary to reduce the velocity of flow in tall stacks.

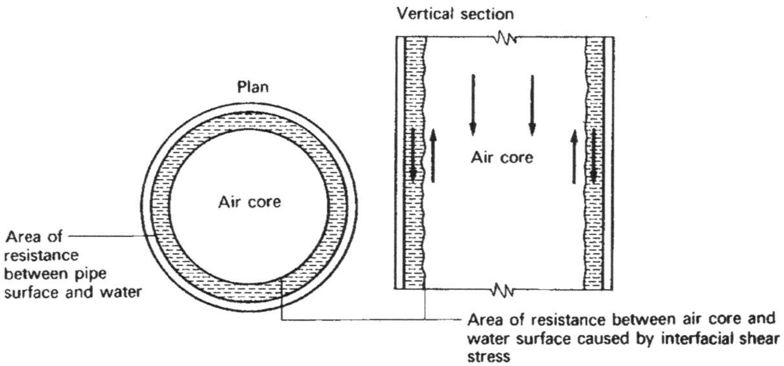

Fig.4.6 Frictional resistances balancing gravitation of water

Capacities of stacks

Owing to the terminal velocity attained in stacks there is a limit to the flow capacity of any given stack. In order to reduce water and air disturbance to a minimum, the upper limit of a stack loading should be about a quarter full. This amount of loading provides room for an air core to flow down the stack during discharges (see Fig. 4.7).

Figure 4.8 shows how the air core is maintained at branch connections to the stack.

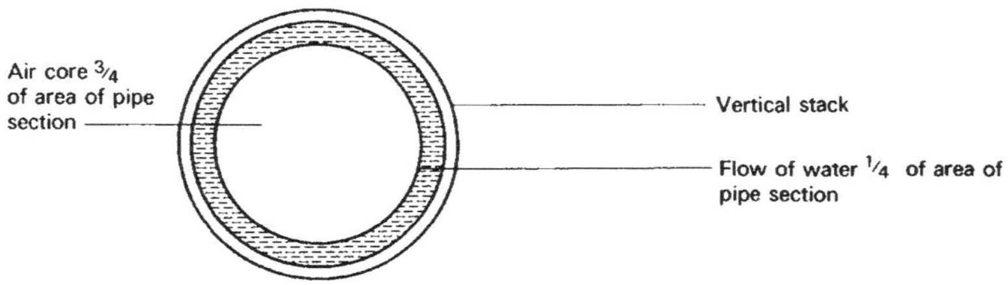

Fig.4.7 Relationship between flow of water and air core

A formula for the sizing of a vertical stack when flowing quarter full is expressed as follows:

$$q = K\,d^{\frac{8}{3}} \text{ or } K\sqrt[3]{d^8}$$

where K = a constant, 0.000 032
q = discharge capacity in litres per second
d = diameter of stack in mm

Example 4.12. *Calculate the diameter of a vertical stack to discharge 2 litres per second when flowing quarter full.*

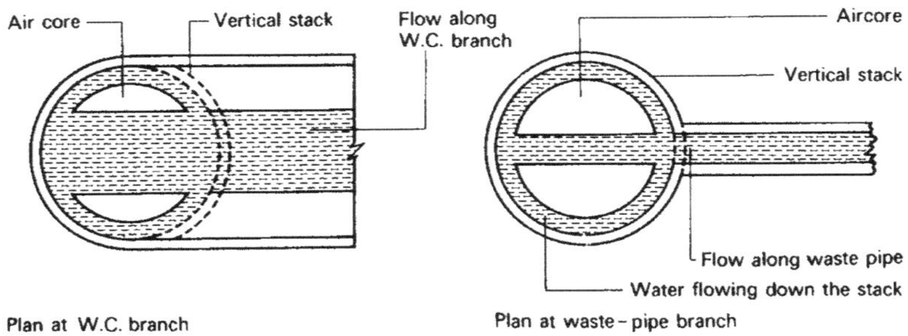

Fig.4.8 Air core at branch connection

$$q = 0.000\,032\,\sqrt[3]{d^8}$$

by transposition $d = \sqrt[8]{\left[\dfrac{q}{0.000\,032}\right]^3}$

$$= \sqrt[8]{\left[\dfrac{2}{0.000\,032}\right]^3}$$

$$= 63 \text{ mm}$$

A 64 mm diameter stack would be suitable.

Discharge unit value method

This method has been described previously for horizontal drains but the method may also be used for vertical discharge stacks using the Table 4.3 to find the sum of the discharge units and then referring to Table 4.6 to find the diameter of the stack.

Table 4.6 Maximum number of discharge units to be allowed on vertical stacks BS 5572 1978

Nominal internal diameter of pipe (mm)	Discharge units
50	10
65	60
75	200 (not more than 1 WC)
90	350
100	750
125	2500
150	5500

Example 4.13. *By use of the discharge unit value method find the diameter of a vertical stack to take the discharges from the following sanitary fittings in a 10 storey office block, 120 W.Cs with 9 litre flush, 120 basins, 20 sinks, and 40 urinals.*

W.Cs	120 × 14 DUs	=	1680
Basins	120 × 3 DUs	=	360
Sinks	20 × 14 DUs	=	280
Urinals	40 × 0.3 DUs	=	12
		Total	2332

By reference to Table 4.6, a 125 mm diameter stack would be suitable.

Diameter of ventilating pipes

The diameters of branch ventilating pipes are given in Table 4.7. Longer lengths of ventilating pipes will, however, have to be larger.

Table 4.7 Diameter of branch ventilating pipes

Diameter of discharge stack (mm)	Diameter of branch ventilating pipe (mm)
Smaller than 75 mm	$\frac{2}{3}D$
75 mm to 150 mm	½D

Sewage disposal

Septic tanks

This is where the primary treatment takes place and sewage should be retained in the tank between 16 and 48 hours. The tank should be sized accordingly and the Code of Practice BS 6297 Small Sewage Treatment Work 1983, gives the following formula for the total capacity of a septic tank where desludging is carried out at not more than 12-monthly intervals:

$$C = 180\,N + 2000$$

where C = capacity of tank in litres
N = number of people served (with a minimum of four)

The Code of Practice recommends that septic tanks which are to serve a population of up to 100 persons should be divided into two compartments. The first compartment should have a length of twice its breadth and should provide two-thirds of the total capacity. The second compartment should be square on plan and sized to hold the remaining one-third of the total capacity. The total length of the tank therefore, ignoring the dividing wall, will be three times the breadth. The depth of the tank should be not less than 1.5 m.

Biological filter

This is where the final treatment of the sewage takes place and where it is purified sufficiently to discharge the effluent into a river, stream or soakaway. It is filled with filter media and its volume should be 1 m^3 per head for up to 10 persons, 0.8 m^3 per head for up to 50 persons and 0.6 m^3 per head for up to 300 persons. The volume should be increased by about 30 per cent when sink grinders are installed. Its depth should be 1.8 m.

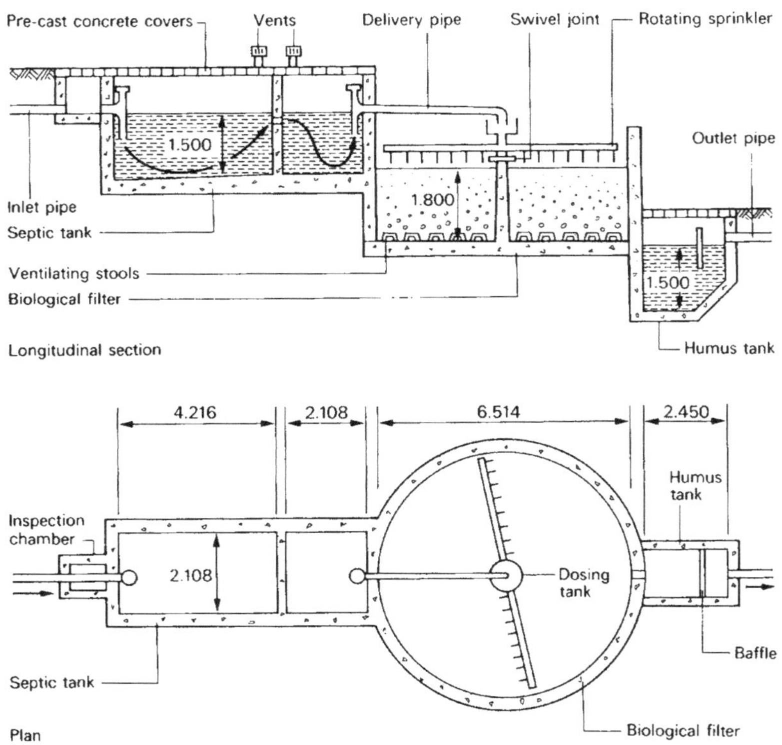

Fig.4.9 Sewage treatment plant for 100 persons

Humus tank

This may be required for larger sewage treatment plants to remove the humus from the effluent of the biological filter. The humus is an inoffensive, earth-like material which, if not removed, may form banks at the river or stream.

The following formula for the capacity of a humus tank for retention times, varying from 24 to 8 hours for a population range 10 to 8,and from 8 to 5 hours for a population range 50 to 300 may be used.

$C = 30N + 1500$

where C = capacity of tank in litres

N = number of people served

Example 4.14. *Calculate the sizes of the septic tank, biological filter and humus tank for a sewage treatment plant to serve 100 persons.*

Dimensions of septic tank

total capacity $= 180\,N + 2000$

$= (180 + 100) + 2000$

$= 20\,000$ litres

$= 20\ m^3$

Internal dimensions of chambers

first chamber volume $= 20 \times \frac{2}{3} = 13.333\ m^3$

length of chamber = twice the breadth, and assuming an average depth of 1.5 m,

volume $= 2\,B \times B \times 1.5$

$= 3\,B^2$

$B = \sqrt{\frac{V}{3}}$

$= \sqrt{\frac{13.333}{3}}$

$= 2.108$

length $= 2.108 \times 2 = 4.216$

internal dimensions of first chamber $= 4.216 \times 2.108 \times 1.500$

internal dimensions of second chamber $= 2.108 \times 2.108 \times 1.500$

Dimensions of biological filter

Assuming a circular filter bed on plan and an average depth of 1.8 m,

volume $= 0.6 \times 100$

$= 60\ m^3$

also volume $= \frac{\pi}{4} D^2 \times 1.8$

$\therefore$ diameter $= \sqrt{\frac{4\,V}{1.8\,\pi}}$

$= \sqrt{\frac{4 \times 60}{1.8 \times 3.142}}$

$= 6.514$

dimensions of filter = 6.514 diameter, and 1.800 deep

Internal dimensions of humus tank

$C = 30N + 1500$

$C = (30 \times 100) + 1.500$

$C = 4500$ litres

volume $= 4.5\ m^3$

Assuming a length of twice the breadth and an average depth of 1.5 m,

$$\text{volume} = 2\,B^2 \times 1.5$$

$$B = \sqrt{\frac{V}{3}}$$

$$= \sqrt{\frac{4.5}{3}}$$

$$= 1.225 \text{ m}$$

$$\text{dimensions} = 2.450 \times 1.225 \times 1.500$$

Questions

(Use Chezy constants = 55 for all questions)

1. Calculate the velocity of flow in m/s and discharging capacity in m^3/s and litre/s of a 150 mm diameter drain flowing half-full and laid at a gradient of 1 in 120.

Answers: 0.9723 m/s; 0.008 594 m^3/s; 8.594 litre/s.

2. Calculate the velocity of flow in m/s and discharging capacity in m^3/s and litre/s of a 100 mm diameter drain flowing full when laid at a gradient of 1 in 50.

Answers: 1.23 m/s; 0.009 66 m^3/s; 9.66 litre/s.

3. Calculate the gradient required for a 150 mm diameter drain flowing half-full when the velocity required is 1 m/s.

Answer: 1 in 113.4

4. Calculate the gradient required for a 300 mm diameter surface-water sewer when flowing full at a velocity of flow of 0.8 m/s.

Answer: 1 in 354.5

5. Calculate the gradient required for a rectangular channel 300 mm wide when the depth of water is to be 150 mm flowing at a velocity of 1.5 m/s.

Answer: 1 in 101

6. Calculate the diameter and gradient for a surface water drain using the following factors:

   (*a*) rainfall intensity 50 mm/h;
   (*b*) area of surface 3000 m^2;
   (*c*) impermeability factor 0.9;
   (*d*) velocity of flow 1.2 m/s;
   (*e*) full-bore discharge.

Answers: 199.458 (use 200 mm diameter drain);
gradient = 1 in 105.

7. A 150 mm diameter drain is to be laid at a gradient of 1 in 180. Calculate the velocity of flow when the drain is flowing half-full.

Answer: 0.7938 m/s (0.8 m/s approx.)

8. Define the following terms: terminal velocity; discharge unit; time of concentration; simultaneous demand factor.

9. Calculate the diameter of pipe and the gradient required to take the run-off from an area of one hectare of asphalt surface, using the following factors:
   (*a*) duration of storm 10 minutes;
   (*b*) impermeability factor 0.875;
   (*c*) time of concentration 20 minutes;
   (*d*) full-bore discharge;
   (*e*) velocity of flow 0.8 m/s.

Answers: 852 mm diameter; 1 in 1007 gradient.

10. By use of the discharge-unit method, find the diameter and gradient for a foul-water drain to take the discharges from 12 W.Cs (9-litre flush), 14 basins, and 6 sinks (assume public use).

Answers: 100 mm diameter; 1 in 45 gradient.

11. Calculate the diameter and gradient for a foul-water sewer to take the discharges from 50 houses. The average number of occupants is assumed to be four per house and the water consumption 225 litres per person, per day. The sewer is to be assumed to run half-full bore at a velocity of 0.8 m/s.

Answers: 125 mm diameter; 1 in 48 gradient.

12. By use of the discharge-unit method, find the diameter of a vertical stack to take the discharges from a three-storey building having the following fittings on each floor: 10 W.C.s (9-litre flush), 14 basins, and 2 sinks.

Answer: 125 mm diameter.

13. Calculate the internal dimensions of the septic tank, filter and humus tank to serve a population of 50 persons. Assume the following factors:
    (*a*) depth of septic tank 1.5 m;
    (*b*) depth of filter 1.8 m;
    (*c*) filter to be circular on plan;
    (*d*) depth of humus tank 1 m with length of twice the breadth.

Answers: (*a*) size of septic tank: first chamber = 3.126 × 1.563 × 1.500;
second chamber = 1.563 × 1.563 × 1.500;
(*b*) size of filter 5.319 diameter, 1.800 depth;
(*d*) size of humus tank 2.450 × 1.225 × 1.000.

Chapter 5

Rainwater pipes and gutters, flow over weirs

Rainwater pipes and gutters

The Building Research Establishment (BRE) has produced a design method for the sizing of rainwater pipes and gutters to match the rainfall intensity (*Digests* 188 and 189).

Rainfall intensity

Roof drainage calculations in the UK can usually be based on rainfall intensity of 75 mm/h, but short storms of a high intensity of 150 mm/h or more do occur and this high value should be used for situations where overflowing might cause serious damage.

It has been found that a storm producing a rainfall intensity of 75 mm/h may occur for 5 minutes once in 4 years, or for 20 minutes once in 50 years. An intensity of 150 mm/h may occur for 3 minutes once in 50 years, or for 4 minutes once in 100 years.

Flow load

The rate of run-off from a roof may be found from the following:

$$\text{rate of run-off } (m^3/s) = \frac{\text{effective area } (m^2) \times \text{rainfall intensity } (m/h)}{3600}$$

Example 5.1. *Calculate the flow load from a roof having an effective area of 50 m^2 when the rainfall intensity is 75 mm/h.*

$$\text{rate of run-off} = \frac{50 \times 0.075}{3600}$$

$$= 0.001\ m^3/s \text{ (approx.)}$$

$$= 1 \text{ litre per second}$$

Allowance for wind (see Fig. 5.1)

In order to allow for the strength of the wind, it is suggested that the angle of descent of wind-driven rain should be taken as one unit horizontal for every two units of descent. This means that the effective area of the roof should be taken as the plan area plus half the elevation area. In Fig. 5.1 the effective area would be:

effective area $= b + \frac{c}{2}$ (m) per m of eaves

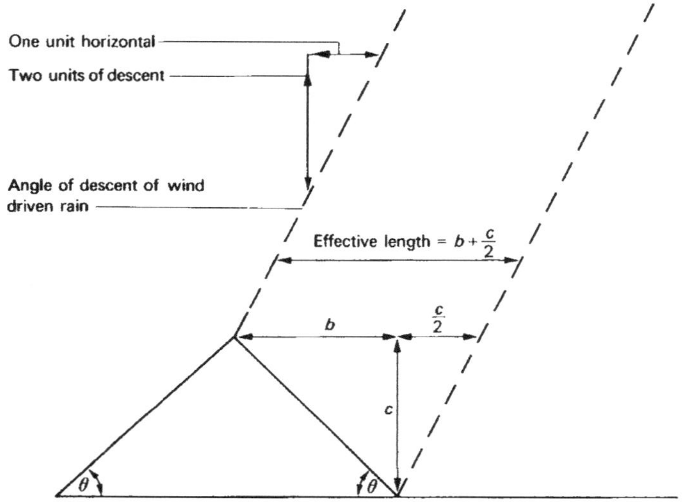

Fig.5.1 Effective length of roof

Run-off from vertical walls (see Fig. 5.2)

If a roof is bounded by vertical walls rising above the roof, some allowance should be made for the additional run-off from these walls. It is not safe to allow for absorption of water by the wall surfaces, even when these are of porous material.

If only one wall is involved, half its area should be added to the roof area to be drained. If the roof is contained within an angle formed by two or more walls, the direction from which these walls present the greatest projected area in elevation should be calculated and half that area added to the roof area to be drained calculated.

Eaves gutters

Each eaves gutter should be designed for the wind direction that would give the maximum rate of flow in the gutter. It is usually convenient in practice to calculate the flow load per metre run of eaves gutter.

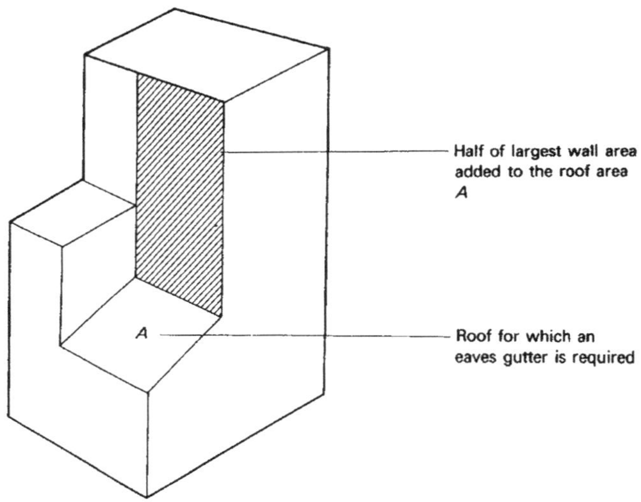

Fig.5.2 Allowance for wall area

Effect of gutter angles

A square angle within 4 m of an outlet will reduce the flow capacity of an eaves gutter. This can be allowed for by multiplying the rate of run-off by one of the following factors:

1. angle within 2 m of the outlet
   sharp cornered × 1.2
   round cornered × 1.1
2. angle within 2–4 m of the outlet
   sharp cornered × 1.1
   round cornered × 1.05

Table 5.1 gives the flow capacities for level gutters, of half-round, segmental and ogee section, with outlet at one end. The flow capacities of gutters of other profiles and sizes may be found from the following formula:

$$\text{flow, litres per second} = \sqrt{\frac{A^3}{B}} \times 10^{-4}$$

where A = area of flow outlet (mm^2)
B = width of water surface at outlet (mm)

Table 5.1

Nominal gutter size (mm)	True half-round gutter (1, 4)	Nominal half-round segmental gutter (2, 3, 4)	Ogee gutter (1)	Ogee gutter (2, 3)
75	0.4	0.3	—	—
100	0.8	0.7	0.9	0.5
115	1.1	0.8	1.4	0.7
125	1.5	1.1	1.7	0.8
150	2.3	1.8	2.6	—

Square PVC gutters are available to similar dimensions.

1. Pressed steel to BS 1091 : 1981
2. Aluminium alloy to BS 2997 : 1980
3. Cast iron to BS 460 : 1981
4. Unplasticised PVC to BS 4576 : 1989 Pt 1

Note: The effective area of roof in square metres to be drained by a level eaves gutter with an outlet at one end when the rainfall intensity is 75 mm/h, can be found by multiplying the flow capacities given in Table 5.1 by 48.

Example 5.2. *Calculate the effective area of roof that may be drained by a level gutter with an outlet at one end when a 100 mm PVC half-round gutter is to be used.*

effective roof area = flow capacity × 48 (Table 5.1)
= 0.8 × 48
= 38.4 m^2

For other rates of rainfall, the areas will be inversely proportional. Gutters fixed with a fall have a greater carrying capacity than when fixed level. Experiments by the Building Research Establishment have shown that an eaves gutter with a fall of 1 in 600 will carry up to 40 per cent more water than when fixed level. The increased flow also helps to prevent silting up of the gutter. Too great a fall looks unsightly and will sometimes allow water falling from the roof to miss the gutter. The rate of flow due to any fall on the gutter may be ignored and if the gutter is assumed to be fixed level the additional size will provide a safety margin.

Sizes of vertical rainwater pipes

The sizes of vertical rainwater pipes appropriate to the gutter sizes given in Table 5.1 are listed in Table 5.2. Figures are given for both sharp- and round-cornered outlets (see Fig. 5.3). Table 5.2 allows also for the position of the outlet in relation to the end of the gutter.

Example 5.3. *Calculate the sizes of eaves gutter and vertical rainwater pipes required to drain a roof 50 m long by 15 m (ridge to eaves) roof pitch 30°.*

Table 5.2 Minimum vertical rainwater pipe sizes (nominal diameter in mm) for various gutter sizes

Half-round gutter size	Sharp- or round-cornered outlet	Outlet at one end of gutter	Outlet not at end of gutter
75	sc	50	50
	rc	50	50
100	sc	63	63
	rc	50	50
115	sc	63	75
	rc	50	63
125	sc	75	89
	rc	63	75
150	sc	89	100
	rc	75	100

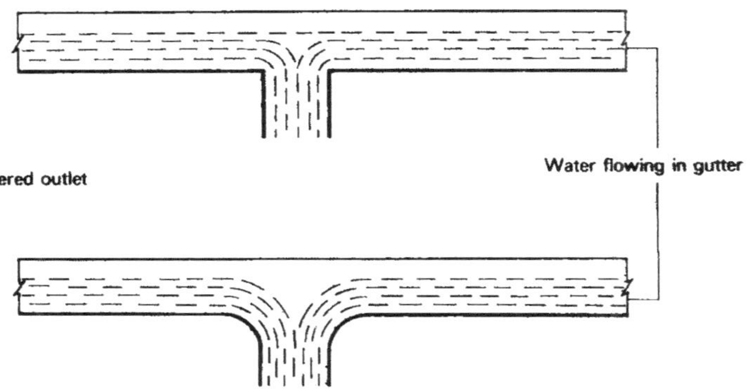

Fig.5.3 Types of gutter outlets

The plan length from ridge to eaves may be found by measurement from drawing or by calculation.

by calculation, plan length $= \cos 30° \times 15$
$= 0.866 \times 15$
$= 12.99$ m

vertical height $= \tan 30° \times 12.99$
$= 0.5774 \times 12.99$
$= 7.5$ m

plan length + half the height $= 12.99 + \dfrac{7.5}{2}$
$= 16.74$ m

effective area of roof per m of eaves $= 16.74\ m^2$

With a rainfall intensity of 75 mm/h the rate of run-off per m of eaves would be:

rate of run-off $= \dfrac{16.74 \times 0.075}{3600}$
$= 0.000\,35\ m^3/s$ per m
$= 0.35$ litre/s per m

From Table 5.1, a 150 mm (nominal) true half-round gutter could cope with a flow from a length of eaves of

$$\frac{2.3}{0.35} = 6.57 \text{ m}$$

Since the gutter outlet receives the flow of water from either side (see Fig. 5.3) and the outlet takes water from only half of the length of gutter to the next outlet, outlets would be required at not more than double the above distance apart.

distance between outlets $= 2 \times 6.57 = 13.14$ m

Figure 5.4 shows a possible solution to the spacing of the rainwater pipes based on this distance.

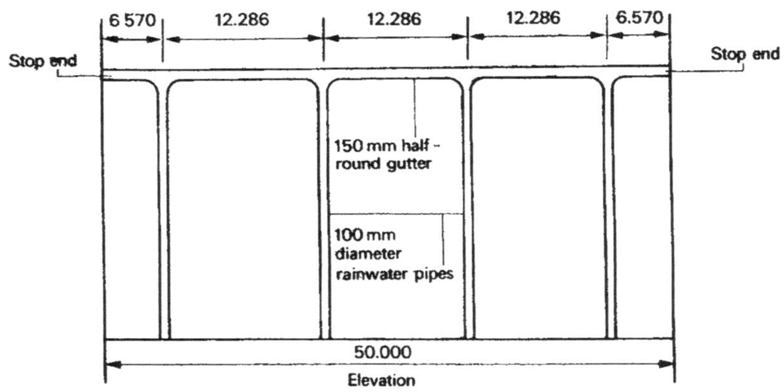

Fig.5.4 Spacing of rainwater pipes

Diameter of pipes

Assuming the use of round-cornered outlets, Table 5.2 shows that 100 mm diameter pipes would be suitable for a 150 mm half-round gutter with the outlets not at the end of the gutter.

Flat roofs

Although the term 'flat roofs' is commonly used, these have a fall of about 1 in 60.

Example 5.4 (see Fig. 5.5) *A flat roof measuring 40 m by 30 m is to be drained by rectangular gutters down each long side. Each gutter is to be provided with an outlet at each end. If a rainfall intensity of 75 mm/h is to be assumed, calculate the size of gutters and the diameter of the outlets and rainwater pipes.*

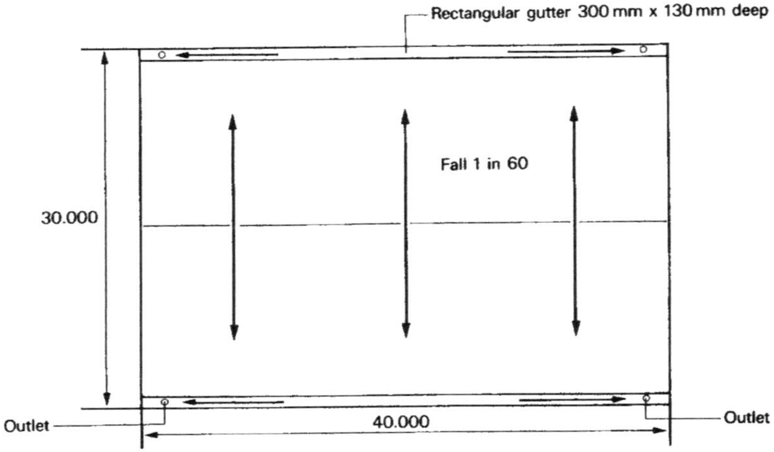

Fig.5.5 Flat-roof drainage

Since each gutter will have a high point at the centre of the gutter, each outlet will take the discharge from one-quarter of the roof, and the gutter will also have to be designed to take this discharge. Therefore,

$$q = \frac{\text{roof area m}^2 \times \text{rainfall intensity m} \times 1000}{4 \times 3600} \text{ litre/s}$$

$$q = \frac{40 \times 30 \times 0.075 \times 1000}{4 \times 3600} \text{ litre/s}$$

$$= 6.25 \text{ litre/s}$$

The Building Research Establishment *Digest* 189, part 2, gives a formula for the flow capacity of a gutter discharging freely, and this is approximately equal to:

$$q = \sqrt{\frac{A^3}{B}} \times 10^{-4} \text{ litre/s}$$

$$\left[\frac{q}{10^{-4}}\right]^2 = \frac{A^3}{B}$$

$$\left[\frac{q}{10^{-4}}\right]^2 \times B = A^3$$

$$A = \sqrt[3]{\left[\frac{q}{10^{-4}}\right]^2 \times B}$$

where A = area of water flow at outlet (mm^2)
B = width of water surface at outlet (mm)

Assume a gutter width of 300 mm. To avoid swirl, the width should be approximately equal to twice the diameter of the outlet (which can be checked later).

$$A = \sqrt[3]{\left[\frac{6.25}{0.0001}\right]^2 \times 300}$$

$$= \sqrt[3]{62\,500^2 \times 300}$$

$$= 10\,520 \text{ mm}^2$$

$$\text{depth of water flow at outlet} = \frac{A}{B} = \frac{10\,520}{300}$$

$$= 35 \text{ mm}$$

For free discharge, the maximum depth of flow in the gutter is twice the depth at the outlet, therefore the maximum flow depth is 2 × 35 = 70 mm. For large gutters, however, a freeboard or height above the water level at the top of the gutter when the water is flowing at its maximum rated depth should be 50 to 60 mm; therefore the final depth of gutter is 70 + 60 = 130 mm.

∴ size of rectangular gutter = 300 mm × 130 mm deep.

Rainwater pipes

Where a gutter discharges directly into a rainwater pipe, the rainwater pipe (or gutter outlet) should be designed to allow free discharge of water from the gutter.

When sizing the outlet to a rainwater pipe leading directly from a gutter, the design head H_d should be taken as the depth of flow at the gutter outlet H_o in Fig. 5.6. If, however, the gutter discharges into a box receiver to which the rainwater pipe connects, the head is the depth of water in the box (see Fig. 5.7).

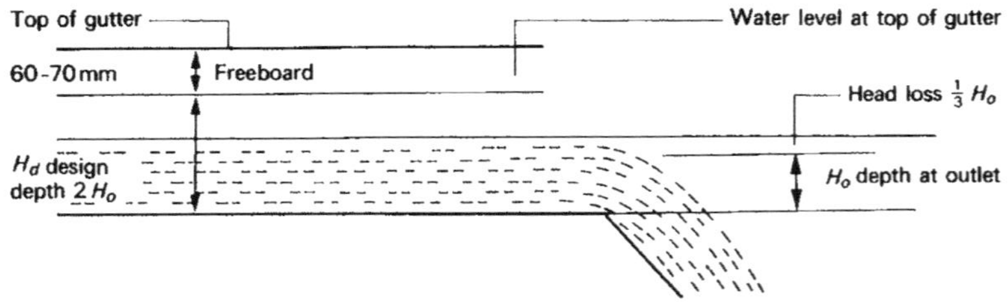

Fig.5.6 Depth of flow from gutter leading directly into a rainwater pipe

The diameter of the rainwater pipe can be reduced to two-thirds of the effective outlet diameter, provided that the transition is gradual over a length of not less than the diameter of the outlet (see Fig. 5.8).

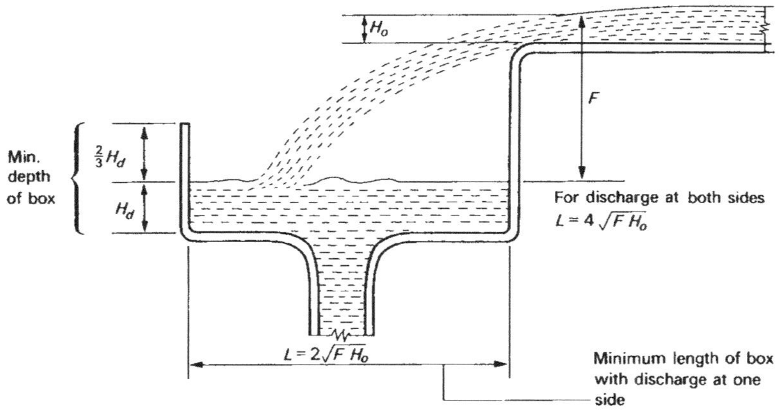

Fig.5.7 Depth of flow from gutter leading to a box receiver

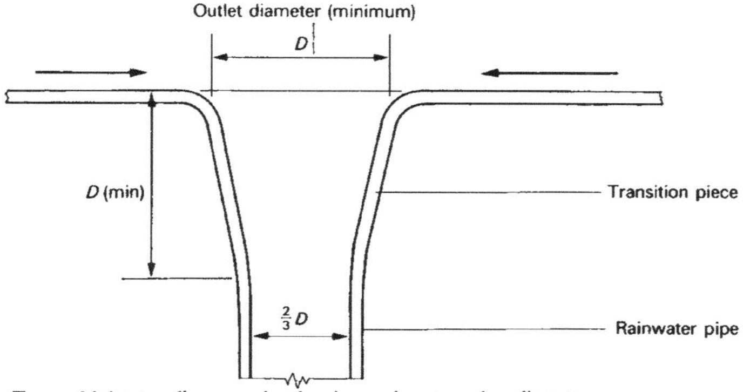

Fig.5.8 Type of inlet to allow a reduction in a rainwater pipe diameter

If the outlet is covered with a grating, the effective discharge capacity of the outlet must be calculated from the unobstructive area of the grating. The discharge capacity of a rainwater pipe depends upon the type of outlet, the smoothness of the pipe bore, the diameter of the outlet, the depth of water over the outlet, and whether the water swirls and forms a vortex about the outlet.

Swirl can be neglected if the centre line of the outlet is at a distance less than its diameter from the nearest vertical side of the gutter or outlet box (see Fig. 5.9). Swirl can also be suppressed by placing a vertical guide vane along the axis of the gutter, over the rainwater pipe outlet.

In the absence of swirl the outlet acts as a weir for values of the ratio H_1/D up to one-third and the discharge capacity of outlet in litres per second is:

$$q = \frac{D}{5000}\sqrt{H_1^3}$$

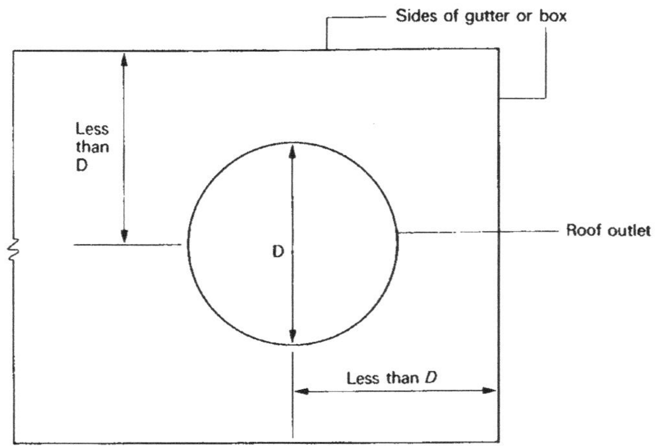

Fig.5.9 Dimensions of gutter or box to avoid swirl at the outlet

where D = diameter of outlet (mm)
H_1 = head of water over outlet (mm)

Where H_1/D is greater than $\frac{1}{3}$, the outlet acts as an orifice and the discharge capacity is:

$$q = \frac{D^2}{15\,000}\sqrt{H_1} \text{ litre/s}$$

Swirl will occur when the centre of the outlet is at a distance greater than its diameter from the nearest vertical side of the gutter or box (unless suppressed by a vane). The outlet then acts as a weir with the same discharge rate as in the absence of swirl for values of H_1/D up to $\frac{1}{4}$; above this, the outlet acts as an orifice but swirling will reduce its discharging capacity to:

$$q = \frac{D^2}{20\,000}\sqrt{H_1} \text{ litre/s}$$

If it is assumed that swirl will be absent, the first formula may be used to find the diameter of the rainwater pipe in example 5.4 and a check made to find if the ratio of H_1/D is less than $\frac{1}{3}$.

$$q = \frac{D}{5000}\sqrt{H_1^3} \text{ litre/s}$$

by transposition

$$\frac{q}{\sqrt{H_1}^3} = \frac{D}{5000}$$

$$\therefore \quad D = \frac{q \times 5000}{\sqrt{H_1^3}}$$

by substitution

$$D = \frac{6.25 \times 5000}{\sqrt{35^3}}$$

$$= \frac{31\,250}{\sqrt{42\,875}}$$

$$= \frac{31\,250}{207}$$

$$= 150.97 \text{ mm}$$

$$= 151 \text{ mm (approx.)}$$

If the effective outlet diameter is reduced gradually over a length of not less than the diameter of the outlet, the diameter of the rainwater pipe may be reduced by two-thirds.

$$\text{diameter of rainwater pipe} = 150.97 \times \frac{2}{3} = 100.64 \text{ mm}$$

$$\text{check on the ratio } H_1/D = \frac{35}{150.97} = 0.232.$$

This is less than $\frac{1}{3}$, and therefore the first formula would be satisfactory. Therefore 4–100 mm diameter pipes would be satisfactory. Some designers may choose 4–125 mm diameter pipes to allow for exceptional storms.

The Institute of Plumbing provide a formula for the sizing of vertical rainwater pipes in the form of:

$$RA = \frac{3.2\,D^{5/2}}{1000}$$

or

$$RA = \frac{3.2\,\sqrt{D^5}}{1000}$$

where RA = roof area in m^2 for a rainfall intensity of 75 mm/h
D = diameter of rainwater pipe in mm

by transposition

$$D = \sqrt[5]{\left[\frac{RA \times 1000}{3.2}\right]^2}$$

$$\text{roof area} = \frac{40 \times 30}{4} = 300 \text{ m}^2$$

$$\therefore \quad D = \sqrt[5]{\left[\frac{300 \times 1000}{3.2}\right]^2}$$

$$= \sqrt[5]{93\,750^2}$$

$$= 97.43 \text{ mm}$$

By use of this formula, it is seen that 4–100 mm diameter pipes would be satisfactory.

The Institute of Plumbing also provide a formula for the sizing of the inlet to the rainwater pipe in the form of:

$$RA = \frac{9}{1000}\,dh^{3/2} \text{ or}$$

$$\frac{9}{1000}\,d\,\sqrt{h^3}$$

where RA = roof area in m^2 for a rainfall intensity of 75 mm/h
d = diameter of outlet in mm
h = head of water over inlet in mm

by transposition

$$d = \frac{300\,000}{\sqrt{h^3} \times 9}$$

$$= \frac{300\,000}{\sqrt{35^3} \times 9}$$

$$= \frac{300\,000}{207 \times 9}$$

$$= 161 \text{ mm} \qquad \text{(BRE 151 mm)}$$

This gives a 10 mm increase in diameter over the BRE method, which is negligible.

Check on gutter width: To avoid swirl at the outlet and a vortex being formed, the width of the gutter should be less than twice the diameter of the outlet: $2 \times 151 = 302$ mm. A 300 mm width is therefore suitable.

Flow of water over weirs

When the sides of an orifice are produced, so that they extend beyond the free surface of the water, it is called a weir. In other words, water flows through an orifice but it flows over a weir. Small weirs are called notches, which usually take the form of thin vertical plates with accurately shaped openings, either rectangular or triangular, and gauging the flow of water.

For example, if the flow of water in a small stream or channel is required, a dam is constructed across the stream and the water allowed to pass through a notch cut in a board or metal plate. The water flowing over a weir or notch is called 'nappe', which corresponds to the jet leaving an orifice.

Ventilation openings

Water flowing over a weir or notch must never be allowed to cling to the downstream face of the plate, but must spring clear. In order to achieve this, ventilation openings must be made in the plate as shown in Fig. 5.10.

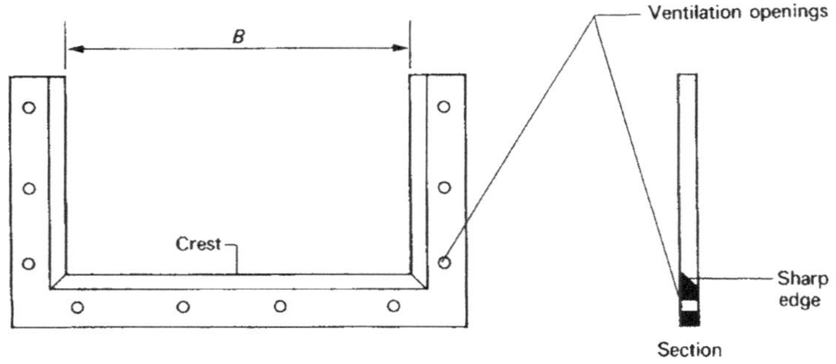

Fig.5.10 Ventilation openings

In the absence of ventilation openings, a pocket of air will be trapped under the nappe and this air will soon be absorbed by the falling water. This will cause a partial vacuum under the nappe which will allow the water to be forced back by the atmospheric pressure acting upon the free water surface until it clings to the downstream face of the weir plate.

Head over crest (*H*) (see Fig. 5.11)

The water level in the upstream channel falls as the stream approaches the weir plate, so that the head over crest is less than the true head which should be measured in still water. The true head, therefore, should be located at a point some distance back from the upstream face of the weir plate. A measuring gauge for this purpose is usually placed upstream at a minimum distance of six times the head over crest.

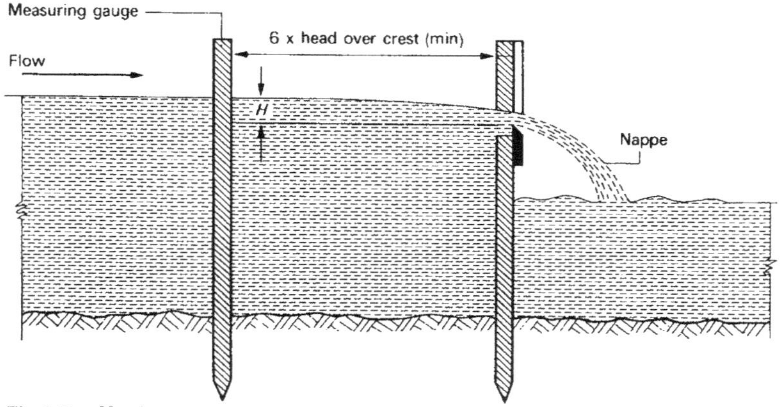

Fig.5.11 Head over crest

Rectangular weirs (see Fig. 5.12)

The edges of the openings may or may not be bevelled. If they are bevelled so as to form a sharp-edged weir, the form of nappe will be different from that formed when the edges are square.

If the sides of the weir are cut square but the edge is sharp, the nappe flows over at full width but the depth will show a contraction due to the sharp-edged crest. This type of weir is called a weir with suppressed end contractions or a suppressed rectangular weir.

If the ends of openings as well as the crest are sharp-edged, the end contractions increase with increasing head. The width of the nappe, however, decreases due to the end contractions. This type of weir is called a fully contracted rectangular weir.

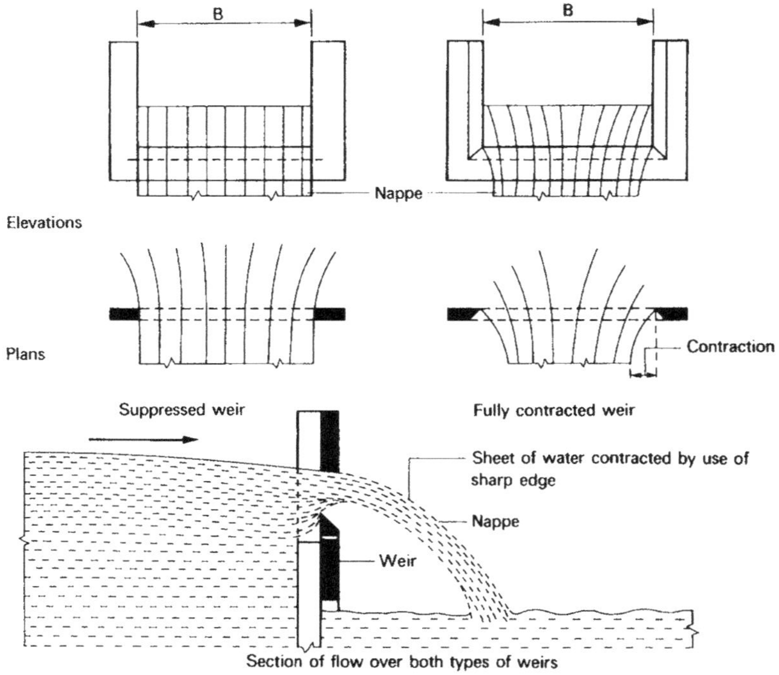

Fig.5.12 Rectangular weirs

Calculations

The formulae used for weirs are as follows:

suppressed weir $Q = 1.82\,B\,\sqrt{H^3}$ (approx.)

fully contracted weir $Q = 1.74\,B\,\sqrt{H^3}$ (approx.)

where Q = flow over weir (m^3/s)
B = breadth of weir (m)
H = head over weir (m)

Example 5.5. *A suppressed rectangular weir is to be used to measure the flow of water in a stream. Calculate the rate of flow in litres per second when the breadth is 600 mm and the head of water over the crest is 150 mm.*

$$Q = 1.82\,B\,\sqrt{H^3}$$
$$= 1.82 \times 0.6 \times \sqrt{0.15^3}$$
$$= 0.063\,34 \text{ m}^3/\text{s}$$
$$= 63.34 \text{ litre/s} \quad \text{(approx.)}$$

Example 5.6. *Calculate the rate of flow over a fully contracted rectangular weir having the same dimensions given in example 5.5.*

$$Q = 1.74\,B\,\sqrt{H^3}$$
$$= 1.74 \times 0.6 \times \sqrt{0.15^3}$$
$$= 0.0606 \text{ m}^3/\text{s}$$
$$= 60.60 \text{ litre/s} \quad \text{(approx.)}$$

Triangular weir (see Fig. 5.13)

For smaller flows of water a triangular weir may be used instead of a rectangular weir. The weir is vee-shaped with the angle of the vee equal to θ. The head of water over the weir is the true head described previously.

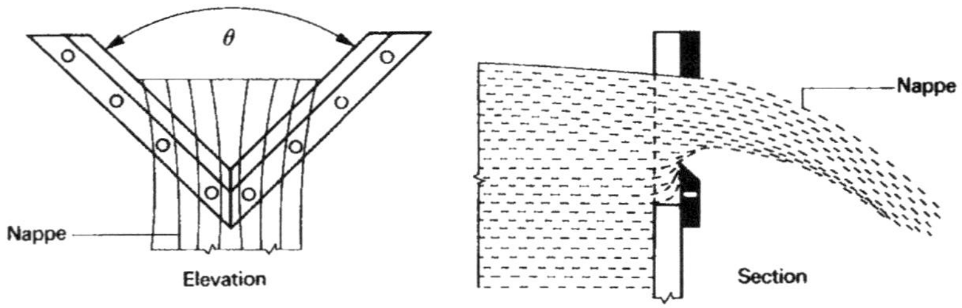

Fig.5.13 Triangular weir

Calculations

The formula used when the angle of vee is 90° is given by:

$$Q = 1.4\,\sqrt{H^5} \quad \text{(approx.)}$$

For an angle other than 90°, the formula becomes:

$$Q = 1.4 \tan\frac{\theta}{2}\,\sqrt{H^5} \quad \text{(approx.)}$$

where Q = flow over weir (m^3/s)
H = head over weir (m)

Example 5.7. *Calculate the rate of flow in litres per second over a 90° triangular weir when the head over the notch is 100 mm.*

$$Q = 1.4\,\sqrt{H^5}$$
$$= 1.4\,\sqrt{0.1^5}$$
$$= 0.004\,43 \text{ m}^3/\text{s}$$
$$= 4.43 \text{ litre/s} \quad \text{(approx.)}$$

Example 5.8. *Calculate the rate of flow over a triangular weir having the same head as that given in example 5.7 but with an angle of 45°.*

$$Q = 1.4 \tan\frac{\theta}{2}\,\sqrt{H^5}$$
$$= 1.4 \times \frac{1}{2}\,\sqrt{0.1^5}$$
$$= 1.4 \times 0.5\,\sqrt{0.1^5}$$
$$= 1.4 \times 0.5 \times 0.003\,162$$
$$= 0.0022 \text{ m}^3/\text{s}$$
$$= 2.2 \text{ litre/s} \quad \text{(approx.)}$$

Questions

1. Calculate the flow load in litres per second from a roof having an effective area of 80 m^2 when the rainfall intensity is 75 mm/h.

Answer: 1.67 litre/s

2. Calculate the effective area of roof that may be drained by a level gutter with an outlet at one end when a 150 mm PVC half-round gutter is to be used.

Answer: 110.4 m^2

3. Calculate the sizes of an eaves gutter and rainwater pipes to drain a roof 30 m long when the plan length from ridge to eaves is 10 m and the vertical height 8 m. Draw a plan of the gutter to show the positions of the vertical rainwater pipes.

Answers:

(*a*)

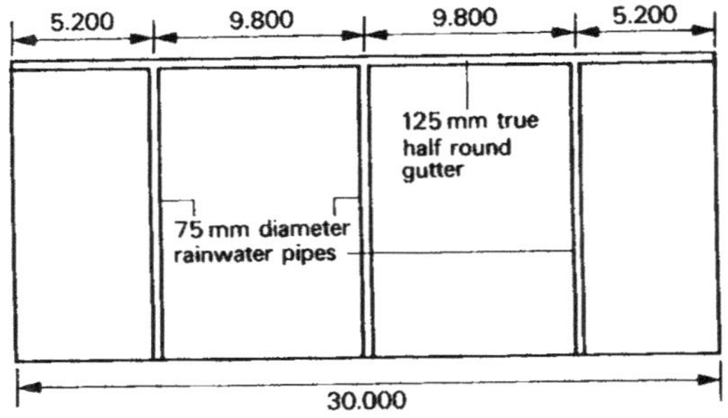

(*b*)

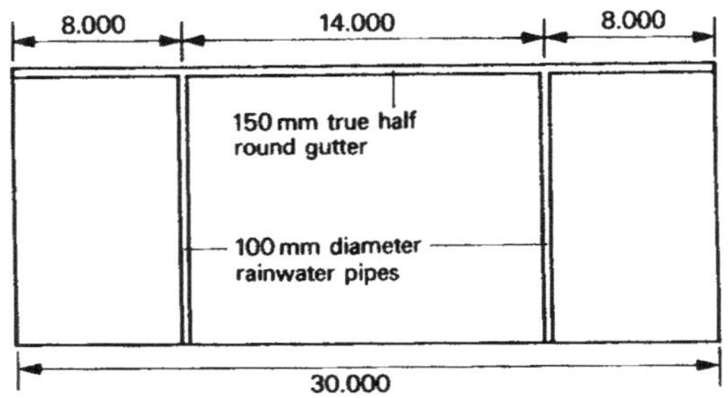

Fig.5.14 Alternative answers to question 3, Chapter 5

4. A flat roof 50 m by 20 m is to be drained by rectangular gutters down each long side. Each gutter is to be provided with an outlet at each end. If a rainfall intensity of 75 mm/h is to be assumed, calculate the sizes of the gutters and the diameter of the rainwater pipes. (Allow freeboard depth of 60 mm.)

Answers: gutter size = 250 mm x 130 mm deep
diameter of pipes = 4–100 mm
outlet diameter = 150 mm

5. A flat roof measuring 20 m by 15 m is to be laid to falls to one rainwater pipe. Assuming a rainfall intensity of 75 mm/h, calculate the diameter of the pipe.

Answer: 100 mm diameter

6. If a gutter discharges from a roof into a box receiver, calculate the minimum length of the box when the box is at the end of the gutter. Height of fall of water 150 mm and depth of water at the discharge point 30 mm.

Answer: 134.2 mm

7. Sketch the type of roof outlet to a rainwater pipe that will provide a flow characteristic sufficient to reduce the pipe diameter by two-thirds. Describe how swirl may be prevented at a roof outlet.

8. Describe the methods used to find the rate of flow in a small stream or channel.

9. A fully contracted rectangular weir has a crest width of 1.200 and a head of water over the crest of 200 mm. Calculate the rate of flow over the weir in litres per second.

Answer: 186.756 litre/s

10. A triangular weir has a vee-angle of 60° and a head of water over the weir of 150 mm. Calculate the rate of flow over the weir in litres per second.

Answer: 10.565 litre/s

Chapter 6

Electrical and gas installations

Electrical installations

Power

The unit of power is the watt (W). If the potential difference between two points of a conductor is 1 V, when carrying a current of 1 A, the power dissipated between the two points is equal to 1 W.

$$W = V\,I \qquad [6.1]$$

where V = potential difference in volts
I = current in amperes

Example 6.1. *An electric cable will carry a current of 10 A when the supply is 240 V. Calculate the power load that may be carried by the cable.*

$$W = V\,I$$
$$= 240 \times 10$$
$$= 2400\text{ W}$$
$$= 2.4\text{ kW}$$

Example 6.2. *A 3 kW electric water heater is connected to a 240 V supply. Calculate the current flowing in the cable when the heater is in use.*

$$W = V\,I$$
$$I = \frac{W}{V}$$
$$= \frac{3000}{240}$$
$$= 12.5\text{ A}$$

Fuses or miniature circuit-breakers

In order to prevent overloading of the cable and possible overheating of the wires, a fuse or miniature circuit-breaker is required which is connected between the live conductor.

In example 6.2 the rating of the fuse or miniature circuit-breaker would normally be 13 A.

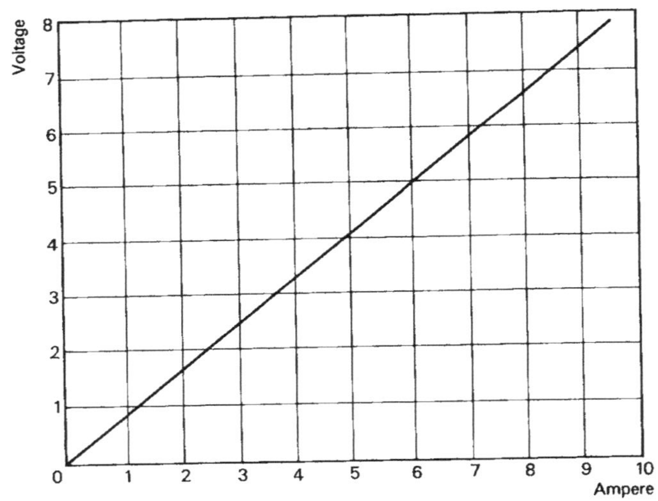

Fig.6.1 Graphical representation of Ohm's law. The gradient gives the resistance ie. $R = \frac{V}{A}$

Ohm's law (See Fig. 6.1)

The current passing through a conductor at constant temperature is proportional to the potential difference between its ends.

$$\text{resistance} = \frac{\text{potential difference}}{\text{current}}$$

$$\text{or } R = \frac{V}{I}\text{ ohms}$$

We can also write

$$V = I\,R \qquad [6.2]$$

Example 6.3. *The potential difference across a resistance through which a current of 30 A is flowing is 50 V. Calculate the resistance.*

$$R = \frac{V}{I} = \frac{50}{30} = 1.667\ \Omega$$

Example 6.4. *What current is flowing through a conductor having a resistance of 50 ohms when the potential difference across the conductor is 240 V?*

$$R = \frac{V}{I}$$

$$I = \frac{V}{R} = \frac{240}{50} = 4.8\ \text{A}$$

Heating effect of current

It is possible to combine the formulas for power and resistance. The formulas [6.1] and [6.2] may be combined as follows:

$$W = V I \quad [6.1]$$

$$V = I R \quad [6.2]$$

Substituting [6.2] in [6.1],

$$W = I R \times I$$

$$W = I^2 \times R \quad [6.3]$$

Also from [6.2],

$$I = \frac{V}{R}$$

and substituting in [6.1]

$$W = V \times \frac{V}{R}$$

$$W = \frac{V^2}{R} \quad [6.4]$$

Note: From formulas [6.3] and [6.4] it can be seen that the power varies either with the square of the current, or the square of the voltage.

Example 6.5. *Calculate the current flowing in and the resistance of the filament of a 150 W tungsten filament lamp (when in use) when connected to a 240 V supply.*

$$W = V I$$

$$I = \frac{W}{V} = \frac{150}{240} = 0.625$$

by Ohm's law $V = I R$

$$R = \frac{V}{I} = \frac{240}{0.625}$$

resistance $= 384\ \Omega$

Check by using formula [6.3]:

$$W = I^2 \times R = 0.625^2 \times 384 = 150$$

Example 6.6. *Calculate the maximum current which can be taken and the resistance of an electric heater element rated at 240 V and 1500 W.*

$$W = V I$$

$$I = \frac{W}{V} = \frac{1500}{240} = 6.25$$

$$R = \frac{V}{I} = \frac{240}{6.25} = 38.4\ \Omega$$

Losses at high voltages

Electricity is transmitted in the National Grid system at high voltages of 132, 275, or 400 kV. The power wasted as heat in these high-voltage transmission cables is very low when compared with the power wasted as heat in low-voltage transmission cables.

Example 6.7. *Calculate the power wasted as internal energy in a cable when 15 kW is transmitted through a cable having a resistance of 0.8 Ω: (a) at 240 V; (b) 275 000 V.*

(*a*) $I = \dfrac{W}{V}$

$$I = \frac{15\ 000}{240}$$
$$= 62.5$$

Power loss in cable (low voltage)

$$W = I^2 R$$
$$= 62.5^2 \times 0.8$$
$$= 3125 \text{ or } 3.125 \text{ kW}$$

(*b*)
$$I = \frac{W}{V}$$
$$= \frac{15\ 000}{275\ 000}$$
$$= 0.0545$$

Power loss in cable (high voltage)

$$W = I^2 R$$
$$= 0.0545^2 \times 0.8$$
$$= 0.002\ 38 \text{ or } 2.38 \text{ mW}$$

We see that the power wasted at high voltages is very low indeed, or almost negligible.

Note: One 400 kV line has three times the power-carrying capacity of one 275 kV line and eighteen times the carrying capacity of one 132 kV line.

Transformers (See Fig. 6.2)

These consist of two distinct windings or coils wound on a laminated iron core. The winding connected on the a.c. supply is known as the primary winding and the secondary winding is connected to the load. An alternating current flows through the primary winding, causing a resultant magnetic flux in the iron core. When this magnetic flux cuts the secondary winding, there is an electromotive force induced in the winding which sends an alternating current through the load.

For a transformer the following relationships apply, if losses are neglected:

$$\frac{\text{primary voltage}}{\text{secondary voltage}} = \frac{\text{primary turns}}{\text{secondary turns}} = \frac{\text{secondary current}}{\text{primary current}}$$

$$\frac{V_1}{V_2} = \frac{N_1}{N_2} = \frac{I_2}{I_1}$$

where
V_1 = primary voltage
V_2 = secondary voltage
N_1 = number of primary turns
N_2 = number of secondary turns
I_1 = primary current
I_2 = secondary current

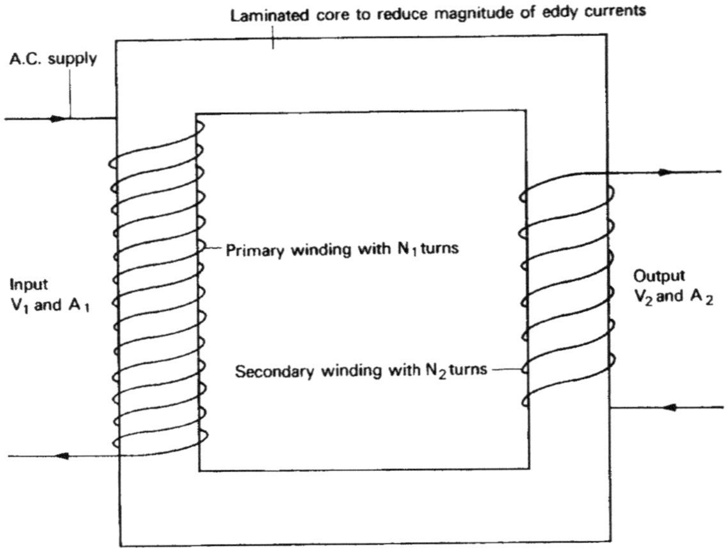

Fig.6.2 Transformer

Example 6.8. *A transformer has 1500 turns in the primary winding and 300 turns in the secondary winding. If the secondary voltage and current are 240 V and 15 A respectively, calculate:* (a) *the primary current;* (b) *the primary voltage.*

(*a*)
$$\frac{N_1}{N_2} = \frac{I_2}{I_1}$$
$$N_1 \times I_1 = N_2 \times I_2$$
$$I_1 = \frac{N_2 \times I_2}{N_1}$$
$$= \frac{300 \times 15}{1500} = 3$$

primary current = 3 amperes

(*b*)
$$\frac{V_1}{V_2} = \frac{N_1}{N_2}$$
$$V_1 \times N_2 = V_2 \times N_1$$
$$V_1 = \frac{V_2 \times N_1}{N_2}$$
$$= \frac{240 \times 1500}{300} = 1200$$

primary voltage = 1200

It will be noticed that the primary high voltage has a lower current than the lower secondary voltage. This results in a lower power loss in the primary circuit as shown in example 6.7.

Area of conductor (see Fig. 6.3)

The resistance offered by a conductor is inversely proportional to its cross-sectional area. In other words, if the cross-sectional area of a conductor is doubled, then resistance is halved, or vice versa.

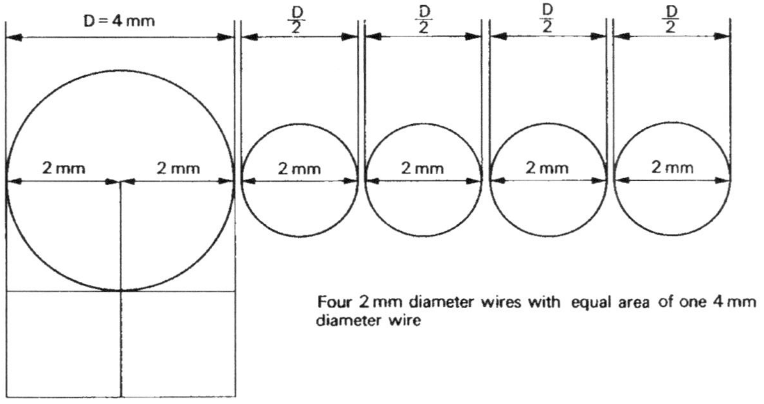

Fig. 6.3 Cross-sectional areas of wires

The areas of circles are to each other as the squares of their diameters, or if a number of small wires are to have the same cross-sectional area as one large wire,

$$N = \frac{D^2}{d^2}$$

where N = number of small-diameter wires
D = diameter of large wire
d = diameter of small wire

Example 6.9. *A certain wire has a diameter of 4 mm and a resistance of 0.5 Ω. What will be the resistance of a 2 mm diameter wire carrying the same values as the 4 mm diameter?*

The number of smaller diameter wires of equal area of the larger diameter wire is given by

$$N = \frac{D^2}{d^2} = \frac{4^2}{2^2} = \frac{16}{4} = 4$$

The large-diameter wire, therefore, is four times greater in area than the small-diameter wire. Therefore, the small-diameter wire will offer four times the resistance of the large-diameter wire.

$4 \times 0.5 = 2\ \Omega$

Length of conductor

If all other values are equal, the resistance of a conductor is proportional to its length.

Example 6.10. *A wire 2 km long has a resistance of 100 Ω. What is the resistance of a piece of this wire 100 m long?*

2000 m offers a resistance of 100 Ω

∴ 100 m offers a resistance of

$$\frac{100}{2000} \times \frac{100}{1} = 5\ \Omega$$

Effect of temperature

With most conductors, a rise in temperature results in an increase in resistance. The resistance of some substances, including carbon, becomes smaller instead of greater as the temperature rises. Certain alloys are also almost unaffected by a change in temperature.

Note: One type of fire-control system operates on the effect of the increase of resistance due to temperature rise. In the system, two copper wires are fixed at high level. One of the wires is insulated against heat and the other left uninsulated. A small electric current is passed through both wires and if a fire occurs, the heat causes a reduction in the flow of current through the uninsulated wire due to the temperature rise. The difference in current flowing through the two wires causes an electric-magnetic valve to open and water is allowed to flow through pipework to open sprinklers.

Root mean square value (r.m.s.) (See Fig. 6.4)

Alternating current and voltage values fluctuate between zero and maximum potential, while direct current values remain steady. In order to obtain the same values for a.c. as are present in d.c., the r.m.s. value is used. The r.m.s. value of a current of 1 A a.c. gives the same heating effect as 1 A d.c.; and similarly for a.c. voltage.

The relationship of r.m.s. value to maximum value a.c. is:

$$\frac{\text{r.m.s. value}}{\text{maximum value}} = 0.707$$

Equivalently,

$$\text{maximum value} = \text{r.m.s. value} \times 1.414$$

Example 6.11.

(*a*) *Calculate the maximum value of a.c. voltage when the r.m.s. value is 240 V.*
(*b*) *Calculate the r.m.s. value of an a.c. current which has a maximum value of 30 A.*

(*a*) maximum value = r.m.s. value × 1.414
= 240 × 1.414
= 339.36 V

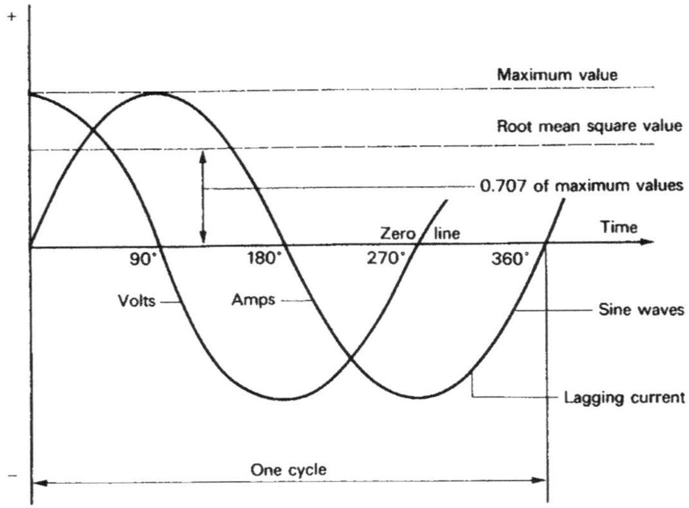

Fig.6.4 Root mean square value

(*b*) r.m.s. value = maximum value × 0.707
= 30 × 0.707
= 21.21 A

Series and parallel circuits

Graphical representation of resistor

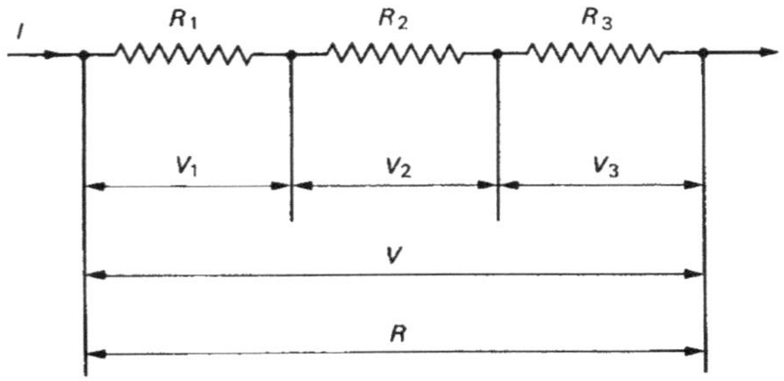

Fig.6.5 Resistors in series

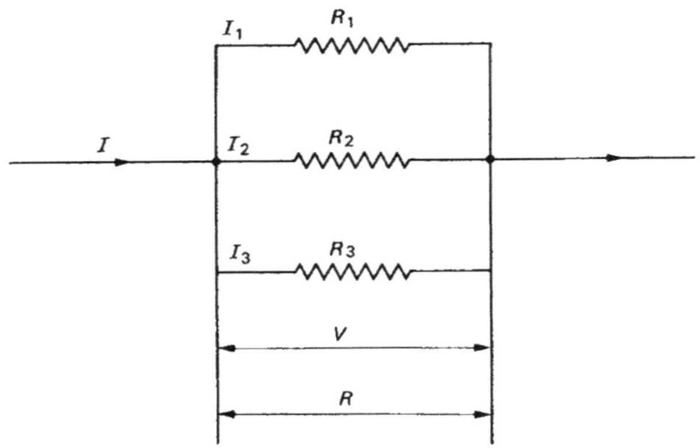

Fig.6.6 Resistors in parallel

Series circuits

If circuits are arranged so that the current passes through resistors consecutively, as shown in Fig. 6.5, the resistors are said to be connected in series.

The same current must pass through all the resistors. If R_1, R_2, R_3, etc. are the resistors, then:

$$V = I R$$
$$V = I R_1 + I R_2 + I R_3$$
$$I R = I R_1 + I R_2 + I R_3$$

The total resistance is therefore the sum of the individual resistances. Dividing throughout by the current,

$$R = R_1 + R_2 + R_3$$

Example 6.12. *What is the total resistance of three resistors connected in series when the individual resistances are 10 Ω, 5 Ω and 3 Ω?*

total resistance = 10 + 5 + 3 = 18 Ω

Parallel circuits

If circuits are arranged side by side with their corresponding ends joined together, as shown in Fig. 6.6, the resistors are said to be connected in parallel.

The same potential difference will apply to each resistor but they will share the same current. Current I will be divided into I_1, I_2 and I_3 through resistances R_1, R_2 and R_3.

$$\text{total current } I = I_1 + I_2 + I_3$$
$$= \frac{V}{R_1} + \frac{V}{R_2} + \frac{V}{R_3}$$
$$\frac{V}{R} = \frac{V}{R_1} + \frac{V}{R_2} + \frac{V}{R_3}$$

Dividing throughout by V,

$$\frac{1}{R} = \frac{1}{R_1} + \frac{1}{R_2} + \frac{1}{R_3}$$

Example 6.13. *Two resistors of 5 and 20 ohms are connected in parallel when the potential difference is 25 volts. Calculate: (a) the total resistance; (b) the total current required; (c) the current through each resistor.*

(*a*)
$$\frac{1}{R} = \frac{1}{R_1} + \frac{1}{R_2}$$
$$= \frac{1}{5} + \frac{1}{20}$$
$$= \frac{4+1}{20}$$
$$= \frac{5}{20}$$
$$\therefore \quad R = \frac{20}{5}$$
$$= 4\ \Omega$$

(*b*)
$$V = I R$$
$$I = \frac{V}{R}$$
$$= \frac{25}{4}$$
$$= 6.25\ \text{A}$$

(*c*)
$$V = I_1 R_1$$
$$I_1 = \frac{V}{R_1} = \frac{25}{5} = 5\ \text{A}$$
$$I_2 = \frac{V}{R_2} = \frac{25}{20} = 1.25\ \text{A}$$

Note: The total current is the same as given in part (*b*).

Power factor

The term 'power factor' applies to alternating current only and is the proportion of current which can be used as energy. A power factor of 0.8 means that 80 per cent of the current supplied is used as energy and the remaining 20 per cent is idle

In an inductive circuit, such as produced by a motor or solenoid, the electromagnetic force opposes the applied voltage which causes the current wave to lag behind the voltage wave. This results in loss of power due to the voltage and current being out of step (see Fig. 6.4).

$$\text{power factor} = \frac{\text{true power}}{\text{apparent power}}$$

True power is measured in watts or kilowatts and apparent power is the product of volts and amperes.

Kilovolt-ampere, kVA

This is the measurement of apparent power in an a.c. supply.

Example 6.14. *Find the apparent power in a 1000 V and 100 A a.c. supply.*

1000 V × 100 A = 1 kV × 100 A = 100 kVA

Example 6.15. *A diesel-electric generator is rated at 500 kVA. If the power factor for the machine is 0.85, find the true power that may be obtained.*

$$\text{kW} = \text{kVA} \times \text{power factor}$$
$$\text{power} = 500 \times 0.85$$
$$= 425\ \text{kW}$$

Example 6.16. *A 2250 W motor on a 240 V single-phase supply is found to take 15 A and 3000 W on full load. Calculate the power factor and percentage efficiency of the motor.*

$$\text{power factor} = \frac{W}{VA} = \frac{3000}{240 \times 15} = 0.833$$
$$\text{efficiency} = \frac{\text{power output}}{\text{power input}} \times \frac{100}{1}$$
$$= \frac{2250}{3000} \times \frac{100}{1} = 75 \text{ per cent}$$

Line and phase voltages

In a four-wire, three-phase supply there are three line conductors carrying a live voltage of 240 each and one neutral return conductor. The voltage between any one of the line conductors and the neutral will provide a single-phase voltage of 240. The voltage between any two line conductors will give a line voltage of 1.73 times the phase voltage.

$$\text{line voltage} = \text{phase voltage} \times \sqrt{3}$$
$$= \text{phase voltage} \times 1.73$$

and

$$\text{phase voltage} = \frac{\text{line voltage}}{1.73}$$

If the phase voltage is 240 the line voltage will be:

$$\text{line voltage} = 240 \times 1.73$$
$$= 415.2 \text{ (usually given as 415 V)}$$

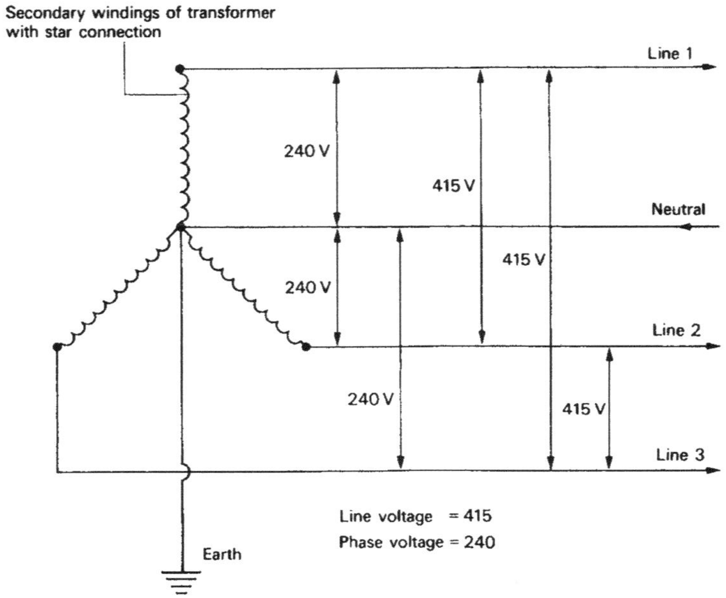

Fig.6.7 Line and phase voltage

Three-phase supplies

For buildings other than dwellings and other small buildings, a three-phase supply is required to meet the higher power load. Electricity boards usually require a three-phase supply when a motor requires 3.73 kW of power or over. It can be shown that three-phase supply is more efficient than single-phase.

Example 6.17. *Calculate the gain in power, expressed as a percentage, of three-phase over single-phase supply, assuming unity power factor and with 240 V and 10 A respectively.*

$$\text{single-phase power} = VI = 240 \times 10 = 2400\ \text{W}$$

$$\text{three-phase power} = VI\sqrt{3} = 240 \times 10 \times 1.73 = 4152\ \text{W}$$

$$\text{gain in power of three-phase} = 4152 - 2400 = 1752\ \text{W}$$

$$\text{gain in power of three-phase, over single-phase expressed as a percentage} = \frac{1752}{2400} \times \frac{100}{1} = 73\ \text{per cent}$$

Therefore for 50 per cent more wire, a three-phase supply provides 73 per cent more power for the same values of voltage and current.

Water heating

It has been found by experiment that the heat energy or the specific heat capacity of water (s.h.c.) is approximately equal to 4.2 kJ/kg $^{\circ}$K. In other words, it requires 4.2 kJ of energy to raise the temperature of 1 kg mass of water through 1 $^{\circ}$C.

Example 6.18. *Calculate the amount of energy required to raise the temperature of 80 kg of water through 60 °C assuming no heat losses.*

$$\text{energy} = \text{s.h.c.} \times \text{kg} \times {}^{\circ}\text{C} = 4.2 \times 80 \times 60$$

$$\text{energy required} = 20\,160\ \text{joules}$$

$$\text{Since kilowatts} = \frac{\text{kilojoules}}{\text{seconds}}$$

the power in kilowatts required, if the water is to be heated in one hour, is:

$$\text{power} = \frac{20\,160}{60 \times 60} = 5.6\ \text{kW}$$

In practice, an allowance will have to be made for the loss of heat from the hot-water storage vessel, even though the vessel may be covered with a good thickness of insulation, say 76 mm. It is not usual to heat stored water by electricity in only one hour. For stored water therefore, we may write:

$$\text{power} = \frac{\text{s.h.c.} \times \text{temperature rise } ({}^{\circ}\text{C}) \times \text{mass of water (kg)} \times 100}{\text{heating time in seconds} \times \text{efficiency (per cent)}}\ \text{kW}$$

Note: If the heat losses are 20 per cent the efficiency will be 80 per cent.

Kilowatt hour

The unit of electricity is 1 kilowatt of electrical power used for 1 hour, or 1 kWh.

Example 6.19. *Find the number of kWhs and the cost when a 3 kW fire is used for 5 hours and the cost of 1 kWh is 8.5p.*

unit = 3 kW × 5 h = 15 kWh
cost = 15 × 8.5p = 127.5p

Energy, work: The base unit of energy or work is the joule, which is equal to 1 N m.

$$\text{since 1 watt} = \frac{1\ \text{N m}}{\text{s}}$$

$$\text{1 watt} = \frac{1\ \text{joule}}{\text{s}}$$

$$\text{or joules} = \text{watts} \times \text{seconds}$$

Therefore, one unit (kWh) of electricity is given by:

$$\begin{aligned}\text{energy} &= 1000 \times 60 \times 60\\ &= 1000 \times 3600\\ &= 3\,600\,000 \text{ joules}\\ &= 3.6 \times 10^6 \text{ joules or } 3.6 \text{ MJ}\end{aligned}$$

Example 6.20. *Calculate the power in kW required to raise the temperature of 45 kg of water from 15 °C to 70 °C in 2 hours when the heat losses are 20 per cent.*

$$\text{power} = \frac{4.2 \times (70 - 15) \times 45 \times 100}{2 \times 3600 \times 80} \text{ kW}$$

$$= 1.8 \text{ kW}$$

A 2 kW electric immersion heater would be required.

Example 6.21. *Calculate the time in hours that it will take to raise the temperature of 136 kg of water from 10 °C to 60 °C by means of a 3 kW immersion heater when the heat losses are 10 per cent.*

$$\text{power} = \frac{\text{s.h.c.} \times \text{kg} \times \text{temperature rise (°C)} \times 100}{\text{heating time in seconds} \times \text{efficiency}} \text{ kW}$$

By transposition,

$$\text{time in seconds} = \frac{\text{s.h.c.} \times \text{kg} \times \text{°C} \times 100}{\text{kW} \times \text{efficiency}}$$

$$= \frac{4.2 \times 136 \times 50 \times 100}{3 \times 90}$$

$$= 10\,577.777$$

$$\text{time in hours} = \frac{10\,577.777}{3600}$$

$$= 2.938$$

It would take approximately 3 hours to heat the water.

If the cost of 1 kWh is 8.5 p, then the cost of heating the water would be:

$$2.938 \times 3 \times 8.5 = 75\text{p (approx.)}$$

Example 6.22. *Calculate the percentage efficiency of an electric water heater having a capacity of 250 litres when its heating element is 6 kW and the time taken to raise the temperature of water from 5°C to 60°C is 3 hours.*

$$\text{power} = \frac{\text{s.h.c.} \times \text{kg} \times \text{temperature rise (°C)} \times 100}{\text{heating up time in seconds} \times \text{efficiency}} \text{ kW}$$

by transposition,

$$\text{efficiency} = \frac{\text{s.h.c.} \times \text{kg} \times \text{temperature rise (°C)} \times 100}{\text{heating up time in seconds} \times \text{kW}}$$

For practical purposes, the mass of 1 litre of water may be taken as 1 kg.

$$\therefore \quad \text{efficiency} = \frac{4.2 \times 250 \times (60 - 5) \times 100}{3 \times 3600 \times 6}$$

$$= 89.12 \text{ per cent}$$

Example 6.23. *Calculate the power in kW required for an immersion heater element when it is required to raise the temperature of 73 litres of water from 10 °C to 60 °C in 2.5 hours and the heat losses are assumed to be 20 per cent.*

$$\text{power} = \frac{\text{s.h.c.} \times \text{kg} \times \text{temperature rise (°C)} \times 100}{\text{heating up time in seconds} \times \text{efficiency}} \text{ kW}$$

$$= \frac{4.2 \times 73 \times 50 \times 100}{2.5 \times 3600 \times 80} \text{ kW}$$

$$= 2.13 \text{ kW}$$

A 2 kW heater would be satisfactory.

Pumping

The power required for pumping is found as follows:

$$\text{power} = \frac{\text{work done}}{\text{time}}$$

$$= \frac{\text{force} \times \text{distance}}{\text{time}}$$

$$= \frac{\text{newtons} \times \text{distance}}{\text{time}}$$

$$\text{force in newtons} = \text{mass} \times 9.81$$

$$\text{power in watts} = \frac{\text{mass} \times 9.81 \times \text{distance}}{\text{time}}$$

Example 6.24. *A centrifugal pumping set is required to raise 5 kg/s of water vertically to a height of 20 m. Allowing 30 per cent of the height for pipe friction, calculate the power in kilowatts required if the efficiency of the equipment is 70 per cent.*

$$\text{effective height} = 20 + (20 \times 0.3) = 26 \text{ m}$$

$$\text{power in watts} = \frac{\text{newtons} \times \text{metres}}{\text{seconds}}$$

$$= \frac{\text{kg} \times 9.81 \times \text{metres}}{\text{seconds}}$$

$$= \frac{5 \times 9.81 \times 26}{1}$$

$$= 1275.3$$

Allowing 70 per cent efficiency,

$$\text{actual power} = 1275.3 \times \frac{100}{70}$$

$$= 1821.857 \text{ watts}$$

Pump laws

When the impeller diameter is constant the following laws apply:

1. The discharge varies directly with the angular velocity of the impeller.
2. The pressure developed varies as the square of the angular velocity of the impeller.
3. The power absorbed varies as the cube of the angular velocity of the impeller.

The laws may be written:

1. $\frac{q_2}{q_1} = \frac{N_2}{N_1}$
2. $\frac{P_2}{P_1} = \frac{(N_2)^2}{(N_1)^2}$
3. $\frac{W_2}{W_1} = \frac{(N_2)^3}{(N_1)^3}$

where q_1 and q_2 = discharge in litres per second
P_1 and P_2 = pressure in kPa
N_1 and N_2 = angular velocity, revolutions per minute
W_1 and W_2 = power absorbed in watts

Example 6.25. *A centrifugal pumping set absorbs 2 kW and discharges 4 litre/s when the angular velocity of the impeller is 800 revolutions per minute. If the angular velocity of the impeller is increased to 1000 revolutions per minute, calculate:* (a) *new discharge,* (b) *new power absorbed.*

(*a*) $\frac{q_2}{q_1} = \frac{N_2}{N_1}$

$$q_2 = \frac{q_1 \times N_2}{N_1} = \frac{4 \times 1000}{800}$$

$$= 5 \text{ litre/s}$$

(*b*) $\frac{W_2}{W_1} = \frac{(N_2)^3}{(N_1)^3}$

$$W_2 = \frac{W_1 \times (N_2)^3}{(N_1)^3} = \frac{2000 \times 1000^3}{800^3}$$

$$= 3906.25 \text{ watts}$$

Example 6.26 (see Fig. 6.8).

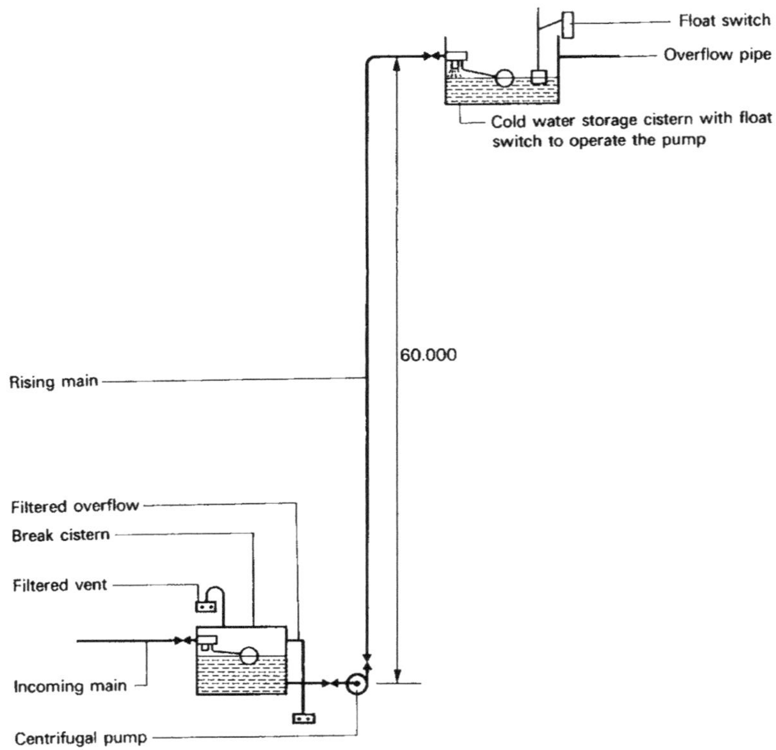

Fig.6.8 Pumping in high-rise building

A cold-water storage cistern is 60 m vertically above a centrifugal pump which is to be used for raising 6000 litres of water from a low-level break cistern every 3 hours. Allowing 30 per cent of the height for pipe friction, calculate the rating of the electric motor if the efficiency of the pumping equipment is 65 per cent.

$$\text{effective height} = 60 + (60 \times 0.3) = 78 \text{ m}$$

$$\text{discharge} = \frac{6000}{3 \times 60 \times 60} = 3.333 \text{ litre/s}$$

$$\text{actual power} = \frac{\text{newtons} \times \text{metres}}{\text{seconds}} \times \frac{100}{\text{efficiency}}$$

$$= \frac{3.333 \times 9.81 \times 78}{1} \times \frac{100}{65}$$

$$= 3923.6075 \text{ watts}$$

required rating of electric motor = 4 kW (approx.)

Electric lifts

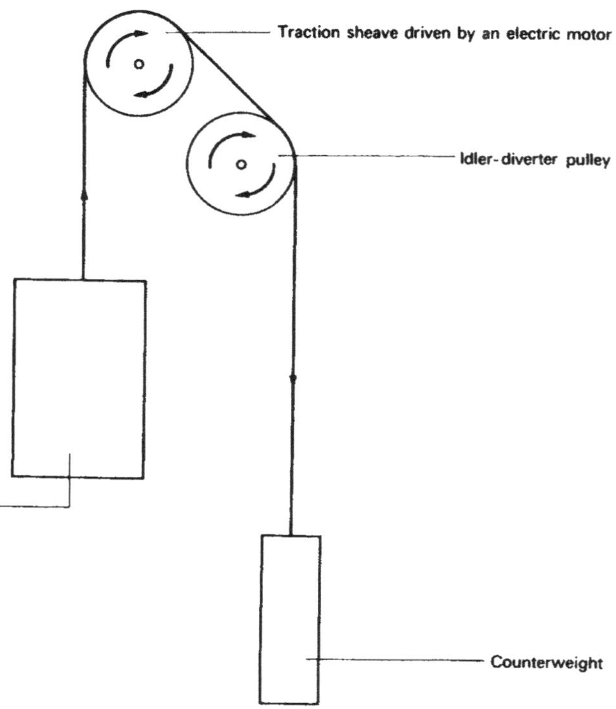

Fig.6.9 Electric lift

The counterweight in an electrically operated lift reduces the load to be moved by the motor to the difference in weight between the car plus its load, and the counterweight plus friction.

$$\text{load to be moved by motor} = \begin{bmatrix}\text{mass of car}\\ \text{plus its load}\end{bmatrix} - \begin{bmatrix}\text{counterweight}\\ \text{plus friction}\end{bmatrix}$$

The counterweight is generally 40–50 per cent of the car plus its load and the friction is generally about 20 per cent of the counterweight.

Example 6.27. ***An electric lift car has a total mass of the car plus its load of 800 kg. If the counterweight has a mass of 50 per cent of the car plus its load, calculate the power in kW required for the electric motor when the car speed is 1.5 m/s and the motor efficiency 60 per cent.***

$$\text{load to be moved by motor} = \begin{bmatrix}\text{mass of car}\\ \text{plus its load}\end{bmatrix} - \begin{bmatrix}\text{counterweight}\\ \text{plus friction}\end{bmatrix}$$

$$\text{counterweight} = 50 \text{ per cent of } 800 = 400 \text{ kg}$$

$$\text{friction} = 20 \text{ per cent of } 400 = 80 \text{ kg}$$

$$\text{mass to be moved by motor} = 800 - [400 + 80] = 320 \text{ kg}$$

$$\text{actual power} = \frac{\text{newtons} \times \text{metres}}{\text{seconds}} \times \frac{100}{\text{efficiency}}$$

$$= \frac{320 \times 9.81 \times 1.5}{1} \times \frac{100}{60}$$

$$= 7848 \text{ watts}$$

$$= 7.85 \text{ kW}$$

Note: In the above calculation of the power, as the speed was 1.5 m/s the distance was put as 1.5 m and the time as 1 s.

Conductor and cable rating

The amount of current which a conductor or cable can carry is limited by the heating effect caused by the resistance to flow of electricity. The maximum permissible current under normal conditions must not be so high that dangerous temperatures may be reached, which could lead to fires. Even with cables inside metal conduits or ducts, or where mineral insulated copper or aluminium sheathed cables are used, although the cables are completely fireproof in themselves, the transmission of heat to other materials in proximity may still lead to fires.

When choosing a cable for a particular duty, it is necessary to take into account not only the maximum current the cable will have to carry, but also the voltage drop that will occur when the current flows.

The 16th edition of the Institute of Electrical Engineers (IEE) Regulations state that the maximum permissible drop in voltage in a conductor shall not exceed 2.5 per cent of the nominal voltage when the conductor is carrying its full load.

The temperature reached by a cable is also affected by the following operating conditions:

1. Whether the cable is surrounded by room air, or is enclosed in a conduit or duct.
2. The proximity to other cables which may cause heat to build up, due to induced currents.
3. The temperature of the ambient or surrounding air.

The British Standards 6004 and 6346 provide tables of cable sizes for various operating conditions (see Table 6.1).

Example 6.28. *A 25 m run of two-core non-armoured PVC insulated copper cable, enclosed in conduit, is required to supply a 10 kW load. If the nominal voltage of the supply is 240, find the size of cable.*

$$\text{current} = \frac{W}{V} = \frac{10\,000}{240} = 41.667 \text{ A}$$

Table 6.1 (Part Table BS 6004 and BS 6346) single-circuit rating and associated voltage drops for single core P.V.C. insulated cables, non-armoured with copper conductors. The ratings tabulated apply where the cable is provided with coarse excess current protection.

Conductor		Bunched and enclosed in conduit or trunking	
		Two cables single phase a.c. or d.c.	
Nominal cross-sectional area (mm^2)	**Number and diameter of wires (mm)**	Current rating (A)	Voltage drop per ampere per metre run
1.0	1/1.13	13	44
1.5	1/1.38	16.5	29
2.5	1/1.78	23	18
4	7/0.85	30	11
6	7/1.04	28	7.3
10	7/1.35	52	4.4

From Table 6.1, a conductor having a cross sectional area of 10 mm^2 will allow a current of 52 A to flow with a voltage drop of 4.4 mV or 0.0044 V per ampere per metre run.

$\therefore$ total voltage drop in cable $= 0.0044 \times 41.667 \times 25$
$= 4.583$ V

check on percentage of nominal voltage:

$$\frac{4.583}{240} \times \frac{100}{1} = 1.909 \text{ per cent}$$

This is below 4 per cent of the nominal voltage and the cable is therefore suitable. suitable.

Gas installations

Flow of gas in pipes

The flow of gas in pipes will depend upon the following factors:

1. the internal diameter of the pipe;
2. the pressure of gas;
3. the type of pipe material used;
4. the relative density of the gas;
5. the length of pipe and resistances of fittings.

Normally the pressure of North Sea gas to the premises is up to 5 kPa, which will be governed down to give a pressure of between 2 to 2.25 kPa in the installation pipework.

Resistances of fittings

In order to find the effective length of pipe, the resistances of fittings must be added to the net length. Table 6.2 gives the additions for fittings.

Table 6.2 Additions to net pipe run (m)

Nominal bore	Elbows	Tees	90° bends
15–25	0.6	0.6	0.3
32–40	1.0	1.0	0.3
50	1.6	1.6	0.3

Rate of flow: Manufacturers of gas appliances usually give the rate of flow in m^3/h.

Table 6.3 gives typical gas consumption at appliances.

Table 6.3 Gas consumption of appliances

Appliance	Consumption (m^3/h)
Refrigerator	0.10
Wash boiler	1.13
Sink water heater	2.30
Instantaneous multi-point water heater	5.70
Cooker	3.70
Warm-air heater	2.30

Table 6.4 gives the discharge rates through steel pipes; it may be used in conjunction with Tables 6.1 and 6.2.

Table 6.4 Discharge rates through steel pipes

Nominal bore (mm)	Length of pipe in metres									
	3	6	9	12	15	18	21	24	27	30
6	0.1									
8	0.1	0.3								
10	1.6	1.3	0.85							
15	3.0	2.1	1.17							
20	6.5	4.5	3.7	3.0	2.8	2.5				
25	13.3	9.4	7.6	6.8	6.0	5.4	5.0	4.8	4.5	
32	26.3	18.4	15.3	13.3	11.9	10.7	10.0	9.3	8.7	8.2
40	26.8	26.0	20.7	18.4	16.4	15.0	13.8	13.0	12.0	11.6
50	73.6	53.8	42.5	36.8	34.0	31.0	28.0	26.3	25.0	23.8

Discharge (m^3/h)

Example 6.29. *Find the internal diameter of a gas pipe to supply one sink water heater, two instantaneous multi-point water heaters, two cookers and* *one refrigerator. The net length of pipe required is 16 m and there will be four 90° bends and five tees in the pipe run.*

Gas consumption (Table 6.3)

sink water heater	=	$2.30\ m^3/h$
instantaneous multi-point water heaters	$= 2 \times 5.70 =$	$11.40\ m^3/h$
cookers	$= 2 \times 3.70 =$	$7.40\ m^3/h$
refrigerator	=	$0.10\ m^3/h$
	total	$21.20\ m^3/h$

Resistance of pipe fittings assuming a 50 mm diameter pipe (see Table 6.2)

90° bends	$= 4 \times 0.6 =$	2.4 m
tees	$= 5 \times 1.6 =$	8.0 m
	total	10.4 m

effective length of pipe = 16 + 10.4 = 26.4 m

Table 6.4 shows that a 50 mm diameter pipe will discharge 25 m^3/h when the length of pipe is 27 m and is therefore adequate.

Pole's formula

The flow of gas in pipes may be found from Pole's formula, which is expressed as follows:

$$q = 0.007755\, d^2 \sqrt{\frac{H\,d}{S\,L}}$$

where q = litres per second
d = diameter of pipe in mm
H = pressure drop in mb
S = relative density of gas (0.58–0.59)
L = length of pipe in m

Calculation of gas consumption

The gas consumed may be found from the following formula:

$$\text{consumption} = \frac{\text{power} \times \text{seconds}}{\text{calorific value}}\ m^3/s$$

The calorific value or heating value of North Sea gas is approximately 37 MJ/m^3, and that of town gas is approximately 19 MJ/m^3.

If the gas appliance is rated in kW this will have to be converted into kJ as follows:

$$\text{power} = \frac{kJ}{\text{seconds}}\ kW$$

$$\therefore \quad kJ = kW \times \text{seconds}$$

Example 6.30. *Calculate the gas consumed in m^3/h by a 25 kW heating boiler when North Sea gas is used.*

$$\text{consumption} = \frac{25 \times 3600}{37\,000} = 2.432\ m^3/h$$

Note: 1 kW = 3.6 MJ/h

Example 6.31. *Calculate the cost of operating a 30 kW gas-fired boiler for 84 hours and the consumption in MJ when 1 kWh costs 2.5p.*

kWh $= 30 \times 84$
$= 2520$
cost $= 2520 \times 2.5p$
$= 6300$
= £63.00

The consumption of gas in MJ may be calculated as follows:

$$kW = \frac{kJ}{s}$$

$$\therefore \quad kJ = kW \times s$$

$$= 30 \times 84 \times 60 \times 60$$

$$MJ = \frac{907\,2000}{1000}$$

$$= 9072$$

Example 6.32. *Calculate the amount of North Sea gas consumed (in m^3) and the cost of heating 136 litres of water from 10°C to 60°C by use of a gas circulator having a power input of 3 kW, when the overall efficiency is 70 per cent. The cost of 1 kWh is 2.5p; the calorific value of gas 38.9 MJ/m^3; the specific heat capacity (s.h.c.) of water is 4.2 kJ/kgK.*

$$\text{power} = \frac{\text{s.h.c.} \times \text{kg} \times {}^\circ\text{C temperature rise} \times 100}{\text{heating time in seconds} \times \text{efficiency}}\ kW$$

$$\text{time in seconds} = \frac{\text{s.h.c.} \times \text{kg} \times {}^\circ\text{C rise} \times 100}{kW \times \text{efficiency}}$$

$$= \frac{4.2 \times 136 \times 50 \times 100}{3 \times 70}$$

$$= 13\,600\ s$$

$$\text{time in hours} = \frac{13\,600}{3600}$$

$$= 3.778\ h$$

MJ per h $= kW \times 3.6$

$= 3 \times 3.6 = 10.8$ MJ

MJ for 3.778 h $= 10.8 \times 3.778$

$= 40.802\ \text{MJ}$

$\text{kWh} = 3.778 \times 3$

$= 11.334$

$\text{cost} = 11.334 \times 2.5$

$= 28.335\text{p}$

$= 28.4\text{p approx.}$

$$\text{also kWh} = \frac{\text{calorific value} \times \text{vol}}{3.6}$$

$$\therefore \quad \text{vol} = \frac{\text{kWh} \times 3.6}{\text{calorific value}}$$

$$= \frac{11.334 \times 3.6}{38.9}$$

$$= 1.0489\ \text{m}^3$$

Example 6.33. *Calculate the operating cost and the volume of gas used in a heating system for one season if the building has a steady design heat loss of 300 kW. Use the following factors:*

(a) Heating season	*30 weeks*
(b) Hours of heating per 24 hours	*10*
(c) Boiler efficiency	*70 per cent*
(d) Load factor	*50 per cent*
(e) Calorific value of gas	*38.9 MJ/m*^{3}
(f) Cost of 1 kWh	*2.5p*

power required allowing
50 per cent load factor = 50 per cent of 300 kW

= 150 kW

time in seconds = $30 \times 7 \times 10 \times 3600$

= 7 560 000 s

kJ = kW × seconds

= $150 \times 7\,560\,000$

= 1 134 000 000

∴ MJ = 1 134 000

$$\text{kWh} = \frac{7\,560\,000}{3600} \times 150$$

$$= 315\,000 \times \frac{100}{70\ \text{per cent}}$$

∴ cost = $450\,000 \times 2.5$

= 1 125 000p

= £11 250

$$\text{vol} = \frac{\text{kWh} \times 3.6}{\text{calorific value}}$$

$$= \frac{315\,000 \times 3.6}{38.9}$$

$$= 29\,151.67\ \text{m}^3$$

Note: The steady design heat loss will be based on an outside air temperature of −1 °C or −2 °C, therefore a load factor is introduced for a heating season in order to take into account the varying outside air temperatures. For a shorter period of, say, one week, the outside air temperature may have a mean of about −1 °C or −2 °C and therefore the load factor will be 100 per cent or unity.

Note: Value added tax at the appropriate rate must be added to the cost of gas and electricity.

Gas pressure

If the gas pressure is given in kilopascals it is sometimes necessary to convert this pressure into millibars or the equivalent reading in millimetres water gauge.

Example 6.34. *The pressure of gas required for an appliance burner is given as 2 kPa. Convert this pressure to millibars pressure and the equivalent reading in millimetres water gauge.*

1 mbar = 100 N/m^2
2 kPa = 200 N/m^2 or 2000 Pa

$$\text{mbars} = \frac{2000}{100}$$

= 20 mbars

pressure in pascals = head in metres × 1000 × 9.81

$$\text{head in metres} = \frac{\text{pressure in pascals}}{1000 \times 9.81}$$

$$= \frac{2000}{1000 \times 9.81}$$

= 0.203 873 5

head of water in millimetres = 203.8735

For practical purposes, a manometer reading of 204 mm water gauge would be satisfactory (see Fig. 6.10).

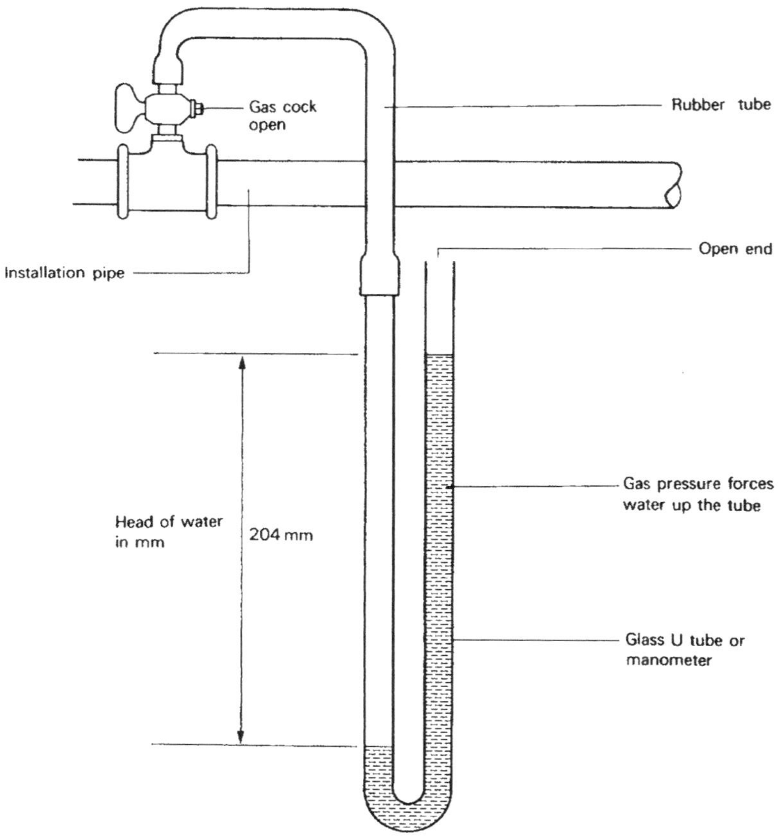

Fig.6.10 Gas pressure

Questions

1. With a supply voltage of 240 V, calculate: (*a*) the current taken by a 2 kW electric immersion heater; (*b*) an 80 W lamp.

Answers: (*a*) 8.333 A; (*b*) 0.333 A

2. Calculate the maximum power load which may be connected to a 30 A ring circuit when the voltage is 240.

Answer: 7200 W or 7.2 kW

3. (*a*) The current in a circuit is 60 A and the resistance 0.2 Ω. Calculate the voltage.
   (*b*) The voltage applied to a circuit is 110 V and the current flowing is 13 A. Calculate the resistance.
   (*c*) If a voltage of 240 V is connected across a resistor of 100 Ω, what current will flow?

Answers: (*a*) 12 V; (*b*) 8.461 Ω; (*c*) 2.4 A

4. Calculate the current flowing in and the resistance of the element of a 5 kW immersion heater (when in use) when connected to a 240 V supply.

Answers: 20.833 A; 11.520 Ω

5. State Ohm's law. Two resistors of 5 Ω and 8 Ω respectively are connected in series, and a potential difference of 12 V is applied to the circuit. Draw the circuit and calculate: (*a*) the total current flowing; (*b*) the voltage across each resistor.

Answers: (*a*) 0.923 A; (*b*) 4.615 V and 7.384 V

6. The electric lighting of a classroom consists of 12 fluorescent lamps, each lamp requiring 100 W at 0.85 power factor. If the supply is 240 V single phase, calculate: (*a*) total power in kW; (*b*) cost of running the lamps for 8 hours when 1 kWh cost 8.5p; (*c*) total kVA; (*d*) current supplied to the circuit.

Answers: (*a*) 1.2 kW; (*b*) 81.6p; (*c*) 1.412; (*d*) 5A

7. A 3 kW motor on a 240 V single-phase supply is found to take 18 A and 3.8 kW on full load. Calculate: (*a*) the power factor; (*b*) the percentage efficiency of the motor.

Answers: (*a*) 0.88; (*b*) 78.947 per cent efficiency

8. Calculate the time taken in hours and the cost of heating 360 kg of water from 5 °C to 60 °C by means of a 8 kW immersion heater when the heat losses are 20 per cent and the cost of 1 kWh is 8.5p.

Answers: 3.609 hours; 245.412p

9. Two 240 V electric lamps, 100 W and 150 W respectively, are connected in parallel to a 240 V supply. Calculate: (*a*) the current flowing through each lamp; (*b*) the resistance of each lamp when in use; (*c*) the total current from the supply.

Answers: (*a*) 0.4166 A and 0.625 A; (*b*) 576 Ω and 384 Ω; (*c*) 1.0416 A

10. A centrifugal pump is required to raise water through a vertical height of 20 m at the rate of 15 kg/s. Allowing for a friction loss on the pipework of 2 m head, calculate the power rating of the motor in kW if the overall efficiency of the pumping unit is 70 per cent.

Answer: 4624.714 W (= 4.6 kW approx.)

11. An electric lift car has a total mass of the car plus its load of 6000 kg. If the counterweight has a mass of 40 per cent of the car plus its load, calculate the power required in kW of the electric motor when the car speed is 0.75 m/s and the motor efficiency 70 per cent.

Answer: 32.8 kW approx.

12. Define: (*a*) kilovolt-amperes; (*b*) power factor; (*c*) root mean square value; (*d*) kilowatts; (*e*) kilowatt hours.

13. A load of 6 kW is to be supplied at 240 volts a.c. through a two-core cable having a resistance per conductor of 0.0025 ohms per metre and the cable length is 10 metres. Find: (*a*) the size of fuse for the circuit; (*b*) the voltage drop over the total length of cable.

Answers: (*a*) 25 A; (*b*) 0.0625 V

14. Calculate the cost of operating a 25 kW gas-fired boiler for 24 hours when 1 kWh costs 2.5p.

Answer: £15.00

15. Calculate the volume of gas consumed (in m^3) and the cost of heating 227 litres of water from 5°C to 60°C by use of a gas storage type water heater having a power input of 5 kW, when the overall efficiency of the heater is 75 per cent. Cost of 1 kWh = 2.5p; calorific value of gas 38.9 M/m^3.

Answer: 1.797 m^3; 48.50p

16. Calculate the operating cost of a heating system for one week if the building has a steady design heat loss of 50 kW. Use the following factors:
    (*a*) Hours of heating per day = 12 hours
    (*b*) Boiler efficiency = 80 per cent
    (*c*) Load factor = 100 per cent
    (*d*) Cost of 1 kWh = 2.5p

Answer: £131.25

17. If a manometer connected to a gas service pipe reads 210 mm, calculate the pressure in millibars.

Answer: 20.6 mbars

Chapter 7

Heating

Heat-loss calculations

Before showing the methods of calculating heat losses from buildings it is necessary to define the terms used.

Definition of terms

1. **Thermal conductivity** (*k*) (see Fig. 7.1) The thermal transmission in unit time through unit area of a slab, or a uniform homogeneous material of unit thickness, when unit difference of temperature is established between its surfaces. The unit is W/m K.

The definition is based on the experimental formula for rate of flow of heat.

$$\frac{Q}{t} = \frac{kA(\theta_1 - \theta_2)}{d}$$

where Q = quantity of heat, measured in joules, passing through the material in a fixed time

t = time in seconds

k = thermal conductivity

θ_1 and θ_2 = temperature of each face of the material (°C)

A = the cross-sectional area of the material through which the heat is passing (m^2)

d = thickness of the material (m)

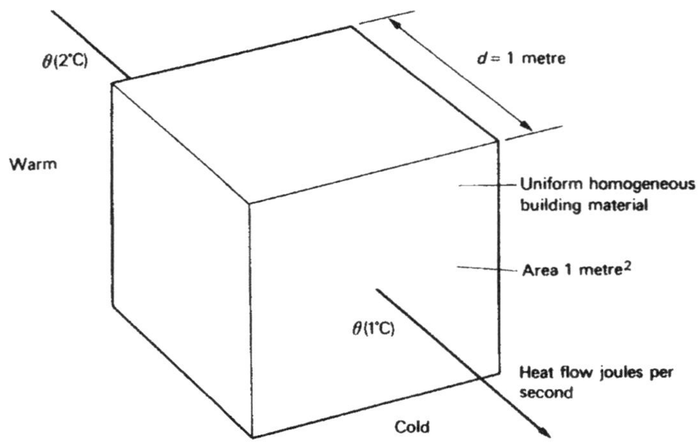

Fig.7.1 Thermal conductivity (k value)

In SI units, thermal conductivity is expressed as heat flow in watts per square metre of surface area, for a temperature difference of 1 °C per metre thickness.

If the units are substituted in the above equation, the units of k will be:

$$k = \frac{\text{J m}}{\text{m}^2 \text{ s K}}$$

since $\text{W} = \frac{\text{J}}{\text{s}}$, this becomes $\frac{\text{W m}}{\text{m}^2 \text{ K}}$

but thickness over area m/m² cancels to 1/m, so that k becomes W/m K.

Table 7.1 gives the k values of common types of building materials.

2. **Thermal resistivity** (r) This is the reciprocal of the thermal conductivity and is used for calculating the total resistance or the conductance of a material. The unit is m K/W (note m²/m cancels to m).

When the thickness of the material is known, its thermal resistance (R) can be calculated: this is the product of thermal resistivity ($1/k$) and thickness and is expressed as m² K/W. The thickness of the material must be in metres.

3. **Thermal conductance** (c) This is the thermal transmission in unit time, through a unit area of a uniform structural component of thickness L per unit of temperature difference between the hot and cold surfaces. The unit is W/m² K.

4. **Thermal resistance** (R) This is the reciprocal of thermal conductance:

$$R = \frac{L}{c}$$

where L is expressed in metres.

Table 7.1 Thermal conductivities (k) of common building materials

Materials	**k value (W/m K)**
Asphalt	1.20
Common brick	1.20
Dense brick	1.47
Aerated concrete	0.14
Structural concrete	1.40
Clinker block	0.05
Cork slab	0.40
Glass wool	0.034
Glass	1.02
Mineral wool	0.037
Gypsum plasterboard	0.15–0.58
Gypsum plaster	0.40
Polyurethane foam	0.020–0.025
Rendering (cement and sand)	0.53
Sandstone	1.30
Limestone	1.50
Clay tiles	0.83
Softwood	0.138
Wood-wool slabs	0.09
Chip board	0.108
Air	0.029
Soil	1.00–1.15
Vermiculite	0.065

For two or more materials in the same structure, the resistance of the individual materials may be added to obtain the resistance in the structure.

$$\text{total resistance} = R_1 + R_2 + R_3 + \ldots$$

5. **Thermal transmission** (U) This is the thermal transmission in unit time through unit area of a given structure. The unit is W/m² K.

The thermal transmission through the structure is obtained by combining the thermal resistances of its components and the adjacent air layers. The thermal transmission therefore is found by adding the thermal resistances and taking the reciprocal.

$$U = \frac{1}{R_{si} + R_{so} + R_1 + R_2 + R_3 + R_a}$$

where U = thermal transmission (W/m² K)
R_{si} = inside surface resistance (m² K/W)
R_{so} = outside surface resistance (m² K/W)
R_1, R_2, R_3 = thermal resistance of structural component (m² K/W)
R_a = resistance of air space (m² K/W)

In computation of U values the thermal resistance L/k is used.

where k = thermal conductivity (W/m K)
L = thickness in metres of a uniform homogeneous material

$$\therefore U = \frac{1}{R_{si} + R_{so} + L_1/k_1 + L_2/k_2 + L_3/k_3 + R_a}$$

Surface-air effects (see Fig. 7.2)

A layer of stationary air is formed on the inside and outside surfaces of the structure. These layers of stationary air act as insulators and must therefore be included in the calculation to find the total resistance of the structure.

The symbols used to denote these surface resistances in the above formula are R_{si} and R_{so}.

Typical values for these resistances (m^2 K/W) are:

horizontal R_{si} = 0.123
horizontal R_{so} = 0.053
upwards R_{si} = 0.110
downwards R_{si} = 0.150
upwards R_{so} = 0.045

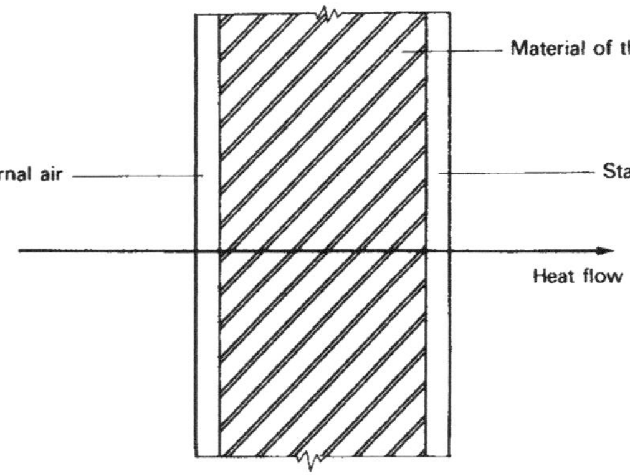

Fig.7.2 Surface-air effects

Resistance of air space

Heat may be transmitted in a cavity by convection, conduction and radiation. If a cavity is less than 20 mm wide it is usual to consider the air as stationary and to deal with it as an extra layer of insulating material. If the cavity is greater than 20 mm wide it is usual to take the resistance as being about 0.18 m^2 K/W.

Calculation of U values

In order to calculate the thermal transmission through a structure, it is preferable to make a sketch of the structure showing the various components and their thicknesses.

Example 7.1. *Figure 7.3 shows a section through a solid brick wall. Using the following data, calculate the thermal transmission (U) for the wall.*

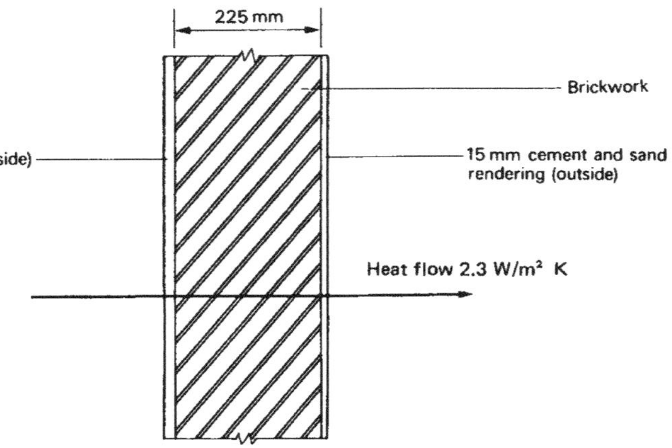

Fig.7.3 Solid brick wall (U value = 2.3 W/m^2 K)

Thermal conductivities (W/m K)

brick	1.20
cement and sand render	0.53
plaster	0.40

Surface resistances (m^2 K/W)

R_{si} internal surface layer	0.123
R_{so} external surface layer	0.053

$$U = \frac{1}{R_{si} + R_{so} + L_1/k_1 + L_2/k_2 + L_3/k_3}$$

$$= \frac{1}{0.123 + 0.053 + 0.019/0.40 + 0.225/1.20 + 0.015/0.53}$$

$$= \frac{1}{0.123 + 0.053 + 0.0475 + 0.1875 + 0.0283}$$

$$= \frac{1}{0.4393}$$

$= 2.276$ W/m^2 K

$= 2.3$ (approx.)

Example 7.2. *Figure 7.4 shows a section through a cavity wall. Using the following data, calculate the thermal transmission (U) for the wall.*

Thermal conductivities (W/m K)

brick	1.20
aerated concrete	0.14
plaster	0.40

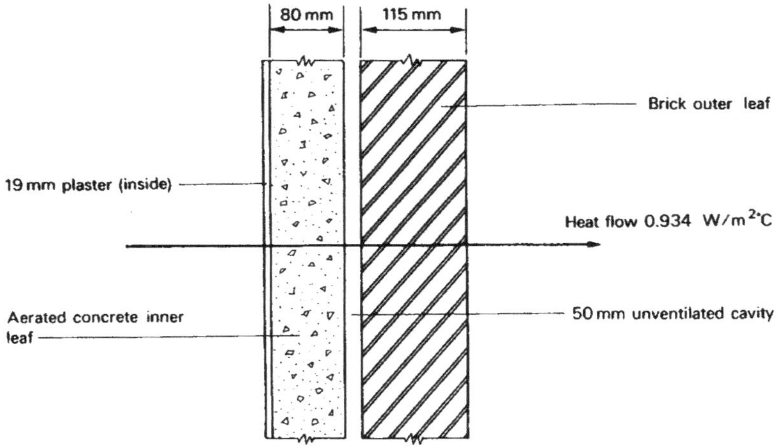

Fig.7.4 Cavity wall (U value = 0.934 W/m^2 °C)

Surface resistances (m^2 K/W)

R_{si} internal surface layer	0.123
R_{so} external surface layer	0.053
R_a air space	0.18

$$U = \frac{1}{R_{si} + R_{so} + L_1/k_1 + L_2/k_2 + L_3/k_3 + R_a}$$

$$= \frac{1}{0.123 + 0.053 + 0.08/0.14 + 0.115/1.20 + 0.019/0.40 + 0.18}$$

$$= \frac{1}{0.123 + 0.053 + 0.571 + 0.096 + 0.0475 + 0.18}$$

$$= \frac{1}{1.0705}$$

$$= 0.934 \text{ W/m}^2 \text{ K}$$

Example 7.3. *If the cavity of the wall given in example 7.2 is filled with polyurethane foam, having a thermal conductivity value of 0.022 W/m K, calculate the U value for the wall and the percentage saving in heat loss. Instead of using 0.18 m² K/W for the resistance of the air space, the resistance of 50 mm thickness of polyurethane foam will be used.*

$$\text{resistance of cavity fill} = \frac{L_4}{k_4} = \frac{0.05}{0.022} = 2.273$$

From the answer given in example 7.2,

$$U = \frac{1}{0.123 + 0.053 + 0.571 + 0.096 + 0.0475 + 2.273}$$

$$= \frac{1}{3.1634}$$

$$= 0.316 \text{ W/m}^2 \text{ K}$$

The percentage saving in heat loss through the wall by filling the cavity with polyurethane foam would be:

difference $= 0.934 - 0.316 = 0.618$ W/m^2 K

saving $= \frac{0.618}{0.934} \times \frac{100}{1} = 66.167$ per cent

With the high cost of fuel, the saving by filling the cavity should ultimately pay for the capital cost of the additional insulation.

Thermal Insulation

Walls.
More heat is lost through the walls of an average building than through other parts of the structure and uninsulated solid walls lose up to 50 per cent more heat than uninsulated cavity walls.

Solid walls may be insulated by fixing aluminium foil-backed plasterboard to timber battens fixed on the inner surface of the wall.

The space between the timber battens may be filled with 12mm to 50mm thick expanded polystyrene slabs.

Alternatively, insulation of the wall may be achieved by fixing resin coating cladding boards, tiles or slates on the external surface. Pebble dash or cement rendering also improves insulation.

Insulation of cavity walls is the most cost effective method of reducing heat losses from a building. Insulation involves filling the cavity within the walls with insulating material such as mineral wool, resin-coated expanded polystyrene beads or urea formaldehyde foam.

The work should be carried out by an approved specialist contractor because not all cavities are suited to insulation. Some buildings may have faults in the wall or may be prone to dampness.

Double glazing.
Correctly installed double glazing halves the heat loss through the windows, improves thermal comfort, reduces condensation and keeps out some external noise. For thermal insulation, a 20mm air space between the panes gives the best result but for sound insulation the air space should be between 100mm and 200mm. Hermetically sealed double glazed windows give very good protection against condensation occuring between the panes. The seal however, is easily damaged and

needs careful handling. If condensation does occur between the panes the seal has been damaged.

Sealed double glazed windows also improves security.

Solid Floors.
These can be insulated by placing insulating slabs above the damp-proof membrane. The floor finish can be placed on top of the insulation.

Hollow timber floors.
These can be insulated by placing insulating boards on top of the joist. The floor boards can be placed on top of the insulation.

Ceilings.
Glass fibre or mineral wool quilt can be placed between the ceiling joists or exfoliated vecmicolite may be poured directly from a bag, between the ceiling joists.

Greenhouse effect.
The reduction in fuel due to insulation reduces the emission of carbon dioxide into the atmosphere and helps to reduce the "greenhouse effect".

Minimum thickness of components

It is sometimes necessary to calculate the minimum thickness of a component forming part of the structure as shown in example 7.4 below.

Example 7.4. ***An external cavity wall is to be constructed so that its U value does not exceed 1 W/m² K. Calculate the minimum thickness of aerated concrete inner leaf of the wall when the brick outer leaf is 115 mm thick; cavity 50 mm wide and internal plastering 19 mm thick.***

Use the following factors:

Thermal conductivities (W/m K)	
brick	1.20
aerated concrete	0.14
plaster	0.40

Surface resistances (m^2 K/W)	
R_{si} internal surface layer	0.123
R_{so} external surface layer	0.053
R_a air space	0.18

$$U = \frac{1}{R_{si} + R_{so} + L_1/k_1 + L_2/k_2 + L_3/k_3 + R_a}$$

$$\frac{1}{U} = R_{si} + R_{so} + L_1/k_1 + L_2/k_2 + L_3/k_3 + R_a$$

Taking the thickness of the inner leaf as L_2 and transposing:

$$1 - R_{si} - R_{so} - \frac{L_1}{k_1} - \frac{L_3}{k_3} - R_a = \frac{L_2}{k_2}$$

$$L_2 = (1 - R_{si} - R_{so} - \frac{L_1}{k_1} - \frac{L_3}{k_3} - R_a) \times k_2$$

$$= (1 - 0.123 - 0.053 - \frac{0.115}{1.2} - \frac{0.019}{0.4} - 0.18) \times 0.14$$

$$= (1 - 0.123 - 0.053 - 0.095\,83 - 0.0475 - 0.18) \times 0.14$$

$$= 0.070\,094 \text{ m}$$

$$= 70 \text{ mm (approx.)}$$

Check:

$$U = \frac{1}{0.123 + 0.053 + 0.0958 + 0.07/0.14 + 0.0475 + 0.18}$$

$$= \frac{1}{0.123 + 0.053 + 0.0958 + 0.5 + 0.0475 + 0.18}$$

$$= \frac{1}{0.9993}$$

$$= 1.000 \text{ W/m}^2 \text{ K}$$

The 70 mm thickness of aerated concrete inner leaf would satisfy the minimum required but an additional saving in fuel would be achieved by a 100 mm thickness.

Heat losses through pitched roofs (see Fig. 7.5)

In order to calculate the U value for pitched roofs with a horizontal ceiling below, the following equation may be used:

$$U = \frac{U_r \times U_c}{U_r + U_c \cos a}$$

where U = the combined thermal transmission for the ceiling and roof (W/m² °C)

U_r = the thermal transmission for the roof (W/m² K)

U_c = the thermal transmission for the ceiling

a = the angle of the pitched roof

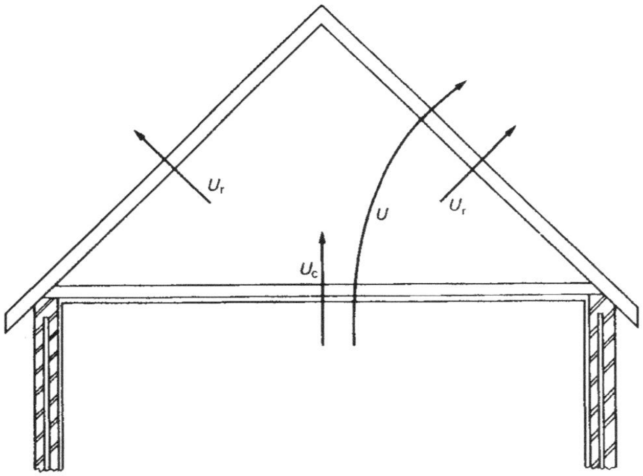

Fig. 7.5 Thermal transmission (U value) for pitched roofs

Example 7.5. *A pitched roof inclined at 40° has a horizontal ceiling below. If* U_r *and* U_c *are found to be 2.50 W/m² K and 0.9 W/m² K respectively, calculate the combined* U *value for the ceiling and roof.*

$$U = \frac{2.5 \times 0.9}{2.5 + (0.9 \times 0.776)}$$

$$= 0.70 \text{ W/m}^2 \text{ K (approx.)}$$

Environmental temperature (t_e)

This is a balanced mean, between the mean radiant temperature and the air temperature. It may be evaluated approximately from the following formula:

$$t_{ei} = \frac{2}{3} t_{ri} + \frac{1}{3} t_{ai}$$

where t_{ri} = mean radiant temperature of all room surfaces (°C)

t_{ai} = inside air temperature (°C)

The approximate value of the mean radiant temperature may be found by totalling the products of the various areas and the surrounding surfaces and dividing this total by the sum of the areas; i.e.,

$$t_{ri} = \frac{a_1 t_1 + a_2 t_2 + a_3 t_3 + \ldots}{a_1 + a_2 + a_3 + \ldots}$$

where

t_1, t_2, t_3 = surface temperature (°C)
a_1, a_2, a_3 = surface areas (m²)

Table 7.3 provides a summary of recommended environmental temperatures.

Table 7.2 Summary of recommended internal (winter) environmental temperatures

	t_{ei} (°C)		t_{ei} (°C)
Art galleries	20	Laboratories	20
Assembly halls	18	Law courts	20
Bars	18	Libraries	20
Canteens	20	*Offices*	
Churches	18	General	20
Factories		Private	20
Sedentary work	19	Police stations	18
Light work	16	Restaurants	18
Heavy work	13	*Hotels*	
Flats and houses		Bedrooms (standard)	22
Living rooms	21	Bedrooms (luxury)	24
Bedrooms	18	Public rooms	21
Bathrooms	22	*Schools and colleges*	
Entrance hall	16	Classrooms	18
Hospitals		Lecture rooms	18
Corridors	16	*Shops*	
Offices	20	Small	18
Operating theatre	18–21	Large	18
Wards	18	Swimming baths	
Sports pavilions	21	Changing rooms	22
Warehouses	16	Bath hall	26

Example 7.6 *Using the following data, calculate the environmental temperature of a room in which the air temperature is 22 °C.*

Fabric	Area (m²)	Surface temperature (°C)
Floor	200	18
Walls	180	20
Windows	60	10
Ceiling	200	16

$$t_{ri} = \frac{(200 \times 18) + (180 \times 20) + (60 \times 10) + (200 \times 16)}{200 + 180 + 60 + 20}$$

$$= \frac{3600 + 3600 + 600 + 3200}{460}$$

$$= \frac{11\,000}{460}$$

$$= 23.9\,°\text{C}$$

The environmental temperature may be found as follows:

$$t_{ei} = \frac{2}{3}\, t_{ri} + \frac{1}{3}\, t_{ai}$$

$$= \left(\frac{2}{3}\right) \times 23.9 + \left(\frac{1}{3}\right) \times 22$$

$$= 15.933 + 7.333$$

$$= 23.266\,^{\circ}\mathrm{C}$$

$$= 23.3\,^{\circ}\mathrm{C}\ \text{(approx.)}$$

The concept of the environmental temperature for heat-loss calculations provides a more accurate assessment of steady-state heat loss through a structure, than does the conventional method that uses internal air temperature, only, as a basis.

The internal environmental temperature is also better than the internal air temperature as an index of the internal comfort of the internal environment. This places the designer in a more favourable position for assessing the internal thermal comfort than is possible by employing air temperature.

The conventional method of using the internal air temperature may be used when the difference between the mean radiant temperature and the air temperature is quite small. This occurs when rooms have little exposure to the outside and the standard of thermal insulation is very high.

The thermal transmission (U) values of common forms of construction may be obtained from various organisations. Table 7.3 provides a list of typical values for normal exposure which may be used as a general guide in computing the heat losses through the structure.

Whether the calculation is carried out by the environmental-temperature concept or the internal air-temperature method, the following formula for the heat loss through the structure is used:

heat loss in watts = area (m^2) × U value × temperature difference

---

Example 7.7. *Compare the heat losses in watts through two walls having* U *values of 1.5 W/m² K and 0.70 W/m² K respectively, when the area of each wall is 200 m². The internal environmental temperature may be taken as 20 °C when the outside air temperature is – 1 °C.*

$$\text{heat loss} = \text{area} \times U \times (t_{ei} - t_{ao})$$

where area = m^2

U = thermal transmission ($W/m^2\,^{\circ}C$)

t_{ei} = environmental temperature inside (°C)

t_{ao} = air temperature outside (°C)

For walls having a U value of 1.50 W/m^2 K,

$$\text{heat loss} = 200 \times 1.50 \times (20 - (-1))$$

$$= 200 \times 1.50 \times 21$$

$$= 6300 \text{ watts}$$

Table 7.3 *U* values for common types of construction

Construction	*U* value (W/m^2 K)
Walls (brickwork)	
220 mm solid brick wall unplastered	2.3
220 mm solid brick wall with 16 mm plaster	2.1
260 mm unventilated cavity wall with 105 mm brick outer and inner leaves, with 16 mm plaster	1.5
260 mm unventilated cavity wall with 105 mm brick outer leaf and 100 mm lightweight concrete block inner leaf and 16 mm plaster	0.96
260 mm cavity wall as above but with 13 mm expanded polystyrene board in cavity	0.70
Flat roofs	
19 mm asphalt on 150 mm solid concrete	3.4
19 mm asphalt on 150 mm hollow tiles	2.2
19 mm asphalt on 13 mm cement and sand screed, 50 mm woodwork slabs on timber joists and aluminium foil backed by 10 mm plasterboard ceiling	0.9
Doors	
25 mm wood or cored	2.4
Partition walls	
75 mm breeze blocks plastered on both sides	2.2
Pitched roofs	
Tiles on battens, sarking felt and plasterboard ceiling	2.44
Tiles on battens, sarking felt and aluminium foil-backed 10 mm plasterboard ceiling	1.50
As above but with 100 mm glass-fibre insulation laid between joists	0.35
Floors	
Solid ground floor with exposed edges 3 m by 3 m	1.47
Solid ground floor with exposed edges 30 m by 30 m	0.26
Solid ground floor with exposed edges 60 m by 60 m	0.11
Intermediate floor on joists with plaster ceiling	1.65
Intermediate floor with 150 mm concrete, 50 mm screed screed and timber flooring	1.87
Suspended timber floor above ground 3 m by 3 m	1.05
Suspended timber floor above ground 30 m by 30 m	0.28
Suspended timber floor above ground 60 m by 60 m	0.16
Glass	
Single glazing	5.6
Double glazing with 20 mm air space	2.9
Triple glazing with 20 mm air space	2.0

For walls having a U value of 0.70 W/m^2 K,

heat loss = $200 \times 0.70 \times 21$

= 2940 watts

difference in heat loss = $6300 - 2940 = 3360$ watts

percentage saving = $\frac{3360}{6300} \times \frac{100}{1}$

= 53.333 per cent

Heat loss by ventilation

The heat lost due to infiltration of external air may be found from the following formula:

$$Q_v = \rho V C (t_{ai} - t_{ao})$$

where Q_v = heat lost (W)

ρ = density of air which may be taken as 1.2 kg/m^3

V = infiltration rate (m^3/s)

C = specific heat capacity of air, which may be taken as 1000 J/kg °K

t_{ai} = inside air temperature (°C)

t_{ao} = outside air temperature (°C)

By introducing the infiltration rate N (air changes per hour) and room volume V (m^3) the formula may be written:

$$Q_v = \frac{\rho N V C}{3600} (t_{ai} - t_{ao})$$

and substituting the above values for ρ and C,

$$Q_v = \frac{1.2 \times N \times V \times 1000}{3600} (t_{ai} - t_{ao})$$

$$= 0.333\, N V (t_{ai} - t_{ao})$$

where Q_v = heat lost in watts
N = rate of air change per hour
V = volume of room m^3
t_{ai} = inside air temperature °C
t_{ao} = outside air temperature °C

Example 7.8. *Calculate the heat lost by ventilation in a room measuring 10 m by 8 m by 3 m when the rate of air change is three per hour and the inside and outside air temperatures are 22 °C and –1 °C respectively.*

$$Q_v = 0.333\, N V (t_{ai} - t_{ao})$$
$$= 0.333 \times 3 \times 240 \times (22 - (-1))$$
$$= 5514.48 \text{ watts}$$

Example 7.9. *A room measuring 8 m by 6 m by 3 m has a heat input of 6 kW when the air temperatures inside and outside are 20 °C and –2 °C respectively. Calculate the rate of air change per hour.*

$$Q_v = 0.333\, N V (t_{ai} - t_{ao})$$

$$N = \frac{Qv}{0.333\, V (t_{ai} - t_{ao})}$$

$$= \frac{6000}{0.333 \times 144 \times 22}$$

$$= 5.687$$

= 6 air changes per hour (approx.)

Note: Some examination questions give the heat capacity of air in J/m^3 K. A common value given is 1340 J/m^3 K.

The formula for the heat lost by ventilation may be written as follows:

$$\text{watts} = \frac{HC \times \text{volume of room } (m^3) \times \text{air change per hr} \times (t_{ai} - t_{ao})}{3600}$$

where HC = heat capacity of air

Example 7.10. *Figure 7.6 shows the plan of a two-bedroomed bungalow. Using the following factors, calculate the heat losses in kW and the required radiator areas in m^2. The bungalow is to be well insulated and the internal air-temperature method of calculating the heat loss through the external walls may therefore be used.*

Factors

1.

Design temperature (°C)		Air change per hr
Dining room	22	2
Lounge	22	2
Kitchen	18	2
Bedrooms	18	1½
Bathroom	22	2
Hall	18	1½
Corridor	18	1½

2.

U values (W/m^2 K)	
Floors	0.30
External walls	0.4
Partition walls	2.20
Windows	2.80
Doors	2.40
Ceilings	0.2

3. External air temperature –1 °C.

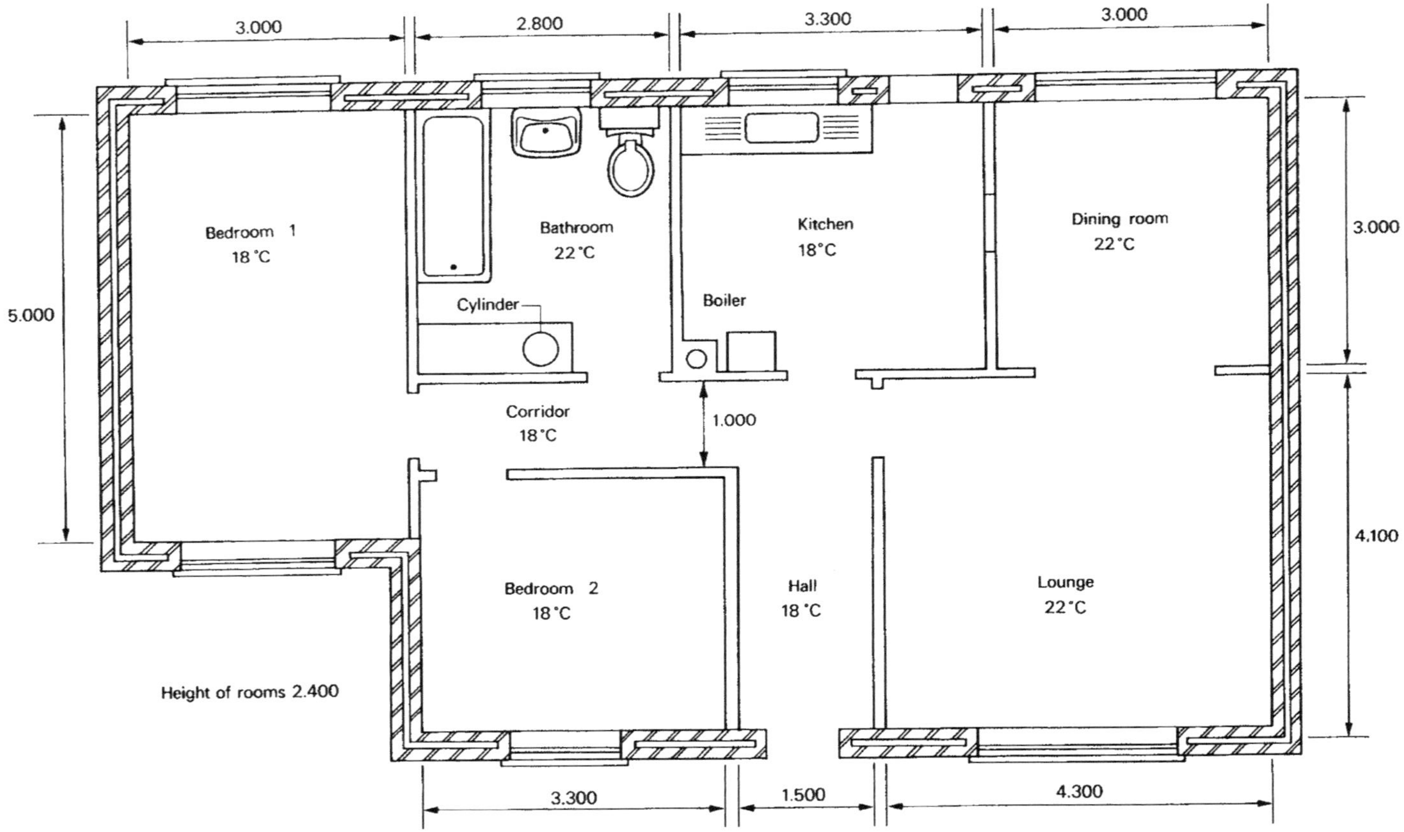

Fig. 7.6 Plan of two-bedroomed bungalow

4. Single-panel steel radiators having a heat emission 0.530 W/m^2.

5. **Window areas (m^2)**

Dining room	2.0 × 1.8
Lounge	1.7 × 1.3
Kitchen	1.2 × 1.0
Bathroom	1.2 × 1.0
Bedroom one (2)	1.7 × 1.2
Bedroom two	1.7 × 1.2

6. Areas of doors (m^2) 1.8 × 0.76

7. **Air volumes**

		m^3
Dining room	3.0 × 3.0 × 2.4	21.60
Lounge	4.3 × 4.1 × 2.4	42.312
Hall	3.2 × 1.5 × 2.4	11.52
Corridor	4.8 × 1.0 × 2.4	11.52
Kitchen	3.3 × 3.0 × 2.4	23.76
Bathroom	3.0 × 2.8 × 2.4	20.16
Bedroom one	5.0 × 3.0 × 2.4	36.00
Bedroom two	3.3 × 3.09 × 2.4	24.47

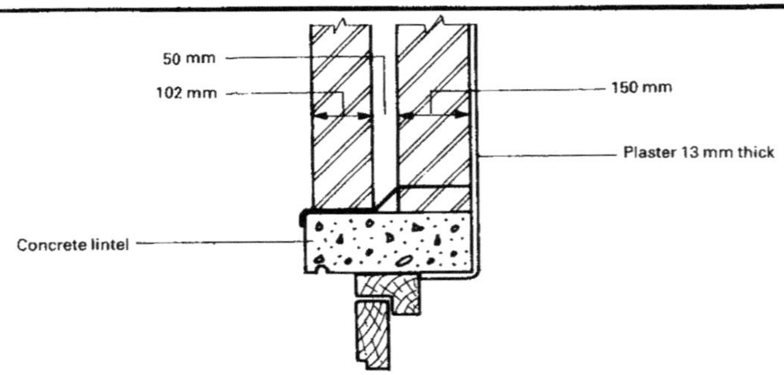

Fig. 7.7 Thermal bridging through concrete lintel

The thermal bridge

A thermal bridge is an area of the building fabric (normally external walling) having a higher thermal transmission than other parts of the fabric. The bridge may be a solid reinforced concrete column bridging a cavity wall, a solid concrete lintel in a cavity wall or a solid window jamb or sill. In constructions, such as steel frames, with insulated panel walls the steel columns will act as a thermal bridge.

The reduction of air temperature may cause the moisture in the air to reach dew point and interstitial condensation may occur in the insulated panels. Condensation may also occur on the surface of the thermal bridge. The edges of the insulated panels may be cooled due to heat from the panels passing to the steel. To reduce the risk of thermal bridging, insulation of the internal bridge surface is required and metal lintels should be filled with insulating foam or mineral wool.

Thermal transmission

To calculate the heat loss through the overall fabric of the building, the higher thermal transmission through the thermal bridge must be found.

Example 7.11 *Calculate the thermal transmission (U value) of the concrete lintel shown in Fig. 7.7. Use the following values:*

Thermal conductivities:

Structural concrete = 1.40 W/mK
Plaster = 0.40 W/mK

Surface resistances:

R_{se} (external surface layer) = 0.12 m^2 K/W

R_{si} (internal surface layer) = 0.08 m^2 K/W

$$U = \frac{1}{R_{se} + R_{si} + \frac{L_1}{k_1} + \frac{L_2}{k_2}}$$

$$U = \frac{1}{0.12 + 0.08 + \frac{0.302}{1.4} + \frac{0.013}{0.4}}$$

$$U = \frac{1}{0.12 + 0.08 + 0.216 + 0.0325}$$

$$U = \frac{1}{0.4485}$$

$$U = 2.2296$$

$$U = 2.23 \text{ W/m}^2\text{ K}$$

Solar heat gains

The heat losses through the building fabric is normally based on equal thermal transmission values for glazed areas on both south and north elevations. A good deal of heat loss may however be saved if windows having larger areas are given a southern aspect; the benefit of this must be taken into account when calculating fabric heat losses. Solar heat gains through glazed areas are instantaneous and this will help to reduce the heating-up time of the room and reduce fuel costs. Too much glazing having a southern aspect should be avoided or overheating will occur during the summer months.

This type of solar heating, which can also warm other parts of the fabric, is known as 'passive'; another type which circulates water through a solar collector is known as 'active'. Both types of solar heating are dealt with in detail in Chapters 11 and 12.

Air infiltration

An air change inside the building is required and some air infiltration is therefore necessary but this must be carefully controlled. Space heating is significantly affected by uncontrolled infiltration of cold air from the outside due to badly fitted doors and windows.

New timber doors and windows are supplied with draught excluding strips. To further reduce heat losses doors may also be thermally insulated. To limit uncontrolled air infiltration the following should also be carried out:

1. sealing of the frames of doors and windows;
2. sealing of suspended timber floors;
3. sealing around edges of skirting boards;
4. sealing around loft hatches;
5. sealing around electrical switches, boxes, sockets and ceiling roses where these project into a cavity.

Global warming

Good insulation and the control of air infiltration will reduce the energy used to heat the building and the costs. This, by contributing to the drive to reduce the national electricity demand, will lead to a reduction in the emission of CO_2. Since the gas CO_2 is a 'greenhouse' gas, its reduced emissions will help against global warming.

Table 7.4 Dining room

Type of structure	Dimensions (m)	Area (m^2)	U value (W/m^2 K)	Temperature difference (K)	Heat loss (W)
External wall	9 × 2.4 deduct 2 × 1.8 (window)	18	0.4	23	165.60
Window	2 × 1.8	3.6	2.8	23	231.84
Partition wall	3 × 2.4	7.2	2.2	4	63.36
Floor	3 × 3	9	0.3	23	62.10
Ceiling	3 × 3	9	0.2	23	41.4
	heat loss through structure			Total	564.30

Dining room

ventilation losses $Q_v = 0.333\ N\ V\ (t_{ai} - t_{ao})$

$= 0.333 \times 2 \times 21.6 \times 23$

$= 330.8688$

total heat losses = 564.3 + 330.87

= 895.17 watts

area of radiator $= \dfrac{895.17}{530}$

$= 1.7\ m^2$ (approx.)

Table 7.5 Lounge

Type of structure	Dimensions (m)	Area (m^2)	U value ($W/m^2\ K$)	Temperature difference (K)	Heat loss (W)
External wall	8.4 × 2.4 deduct 1.7 × 1.3	17.95	0.4	23	165.14
Window	1.7 × 1.3	2.21	2.8	23	142.324
Partition wall	4.1 × 2.4 deduct 1.8 × 0.76	8.472	2.2	4	74.554
Door	1.8 × 0.76	1.368	2.4	4	13.133
Floor	4.3 × 4.1	17.63	0.3	23	121.647
Ceiling	4.3 × 4.1	17.63	0.2	23	81.098
	heat loss through structure			Total	597.896

Lounge

ventilation losses

$Q_v = 0.333\ N\ V\ (t_{ai} - t_{ao})$

$= 0.333 \times 2 \times 42.312 \times 23$

$= 648.135$

total heat losses = 597.896 + 648.135

= 1246.031 watts

area of radiator $= \dfrac{1246.031}{530}$

$= 2.4\ m^2$ (approx.)

Hall and corridor

In order to find the effective heat loss through the structure, the heat gains must be deducted from the heat losses.

Table 7.6 Hall and corridor (heat loss)

Type of structure	Dimensions (m)	Area (m^2)	U value ($W/m^2\ K$)	Temperature difference (K)	Heat loss (W)
External wall	1.5 × 2.4 deduct 1.8 × 0.76	2.232	0.4	19	16.96
Door	1.8 × 0.76	1.368	2.4	19	62.380
Floor	(4.1 × 1.5) add	9.45	0.3	19	53.865
Ceiling	3.3 × 1.0	2.85	0.2	19	10.83
	heat loss through structure			Total	144.035

Table 7.7 Hall and corridor (heat gains)

Type of structure	Dimensions (m)	Area (m^2)	U value ($W/m^2\ K$)	Temperature difference (K)	Heat loss (W)
Partition	(2.8 + 4.1) × 2.4 deduct 2 doors 1.8 × 0.76	13.824	2.2	4	121.651
Doors	2 × 1.8 × 0.76	2.736	2.4	4	26.265
	heat gains through structure			Total	147.916

effective heat loss = 144.035 – 142.916

= 1.119 watts

ventilation losses

$Q_v = 0.333\ N\ V\ (t_{ai} - t_{ao})$

$= 0.333 \times 1.5 \times (2 \times 11.52) \times (18-(-1))$

$= 0.333 \times 1.5 \times 23.04 \times 19$

$= 218.66$

total heat losses = 1.119 + 218.66

= 219.779 watts

area of radiator $= \dfrac{219.779}{530}$

= 0.415

$= 0.5\ m^2$ (approx.)

Table 7.8 Bathroom (heat losses)

Type of structure	Dimensions (m)	Area (m^2)	U values (W/m^2 K)	Temperature difference (K)	Heat loss (W)
External wall	2.8 × 2.4 deduct 1.2 × 1.0	5.52	0.4	23	50.784
Window	1.2 × 1.0	1.2	2.8	23	77.280
Partition walls	11.6 × 2.4 deduct 7.8 × 0.76	26.472	2.2	4	232.954
Door	1.8 × 0.76	1.368	2.4	4	13.133
Floor	3.0 × 2.8	8.4	0.3	23	57.96
Ceiling	3.0 × 2.8	8.4	0.2	23	38.64
	heat loss through structure			Total	412.791

Bathroom

ventilation losses

$$Q_v = 0.333\,N\,V\,(t_{ai} - t_{ao})$$
$$= 0.333 \times 2 \times 20.16 \times 23$$
$$= 308.80$$

total heat losses = 412.791 + 308.8

= 721.591 watts

$$\text{area of radiator} = \frac{721.591}{530}$$
$$= 1.36$$
$$= 1.4\ m^2\ \text{(approx.)}$$

Table 7.9 Bedroom 1

Type of structure	Dimensions (m)	Area (m^2)	U value (W/m^2 K)	Temperature difference (K)	Heat loss (W)
External walls	11 × 2.4 deduct 2 (1.7 × 1.2)	22.32	0.4	19	169.632
Windows	2 (1.7 × 1.2)	4.08	2.8	19	217.056
Floor	5 × 3	15.0	0.4	19	85.500
Ceiling	5 × 3	15.0	0.2	19	57.000
					529.188
Heat gain					
Partition wall	3 × 2.4	7.2	2.2	4	63.360
	heat loss through structure			Total	465.828

Bedroom 1

ventilation losses

$$Q_v = 0.333\,N\,V\,(t_{ai} - t_{ao})$$
$$= 0.333 \times 1.5 \times 36 \times 19$$
$$= 341.658$$

total heat losses = 465.828 + 341.658

= 807.486 watts

$$\text{area of radiators} = \frac{807.486}{530}$$
$$= 1.577$$

∴ Use two radiators, 0.8 m^2 each.

Kitchen

ventilation losses

$$Q_v = 0.333\,N\,V\,(t_{ai} - t_{ao})$$
$$= 0.333 \times 2 \times 23.76 \times 19$$
$$= 300.659$$

total heat loss = 300.659 + 70.865

= 371.524 watts

The heat losses from the boiler would adequately cover the above heat loss and therefore a radiator in the kitchen is not normally required. The heat loss from the kitchen, however, must be added to the heat losses from the other rooms in order to find the boiler power.

Table 7.10 Kitchen

Type of structure	Dimensions (m)	Area (m^2)	U value (W/m^2 K)	Temperature difference (K)	Heat loss (W)
External walls	3.3 × 2.4 deduct 1.2 × 1.0 + 1.8 × 0.76	5.352	0.4	19	40.675
Window	1.2 × 1.0	1.20	2.8	19	6.384
Door	1.8 × 0.76	1.368	2.4	19	6.238
Floor	3.3 × 3.0	9.9	0.3	19	56.430
Ceiling	3.3 × 3.0	9.9	0.2	19	37.62
					260.945
Heat gain					
Partition walls	9.0 × 2.4	21.6	2.2	4	190.080
	heat gain through structure			Total	70.865

Table 7.11 Bedroom 2

Type of structure	Dimensions (m)	Area (m^2)	U value (W/m^2 K)	Temperature difference (K)	Heat loss (W)
External walls	(3.3 + 1.9) × 2.4 deduct 1.7 × 1.2	10.44	0.4	19	79.344
Window	1.7 × 1.2	2.04	2.8	19	108.528
Floor	3.3 × 3.09	10.197	0.3	19	58.123
Ceiling	3.3 × 3.09	10.197	0.2	19	38.749
	heat loss through structure			Total	284.744

Bedroom 2

ventilation losses

$$Q_v = 0.333\ N\ V\ (t_{ai} - t_{ao})$$
$$= 0.333 \times 1.5 \times 24.47 \times 19$$
$$= 232.233$$

total heat losses = 284.744 + 232.233

= 516.977 watts

$$\text{area of radiator} = \frac{516.977}{530}$$
$$= 0.975$$
$$= 1\ m^2\ \text{(approx.)}$$

Total heat losses (watts)

Dining room	895.170
Lounge	1246.031
Hall and corridor	219.779
Bathroom	721.950
Bedroom 1	807.486
Bedroom 2	516.977
Kitchen	371.524
	4778.917

Boiler power

To this heat loss, an allowance of 25 per cent may be added for heat losses from pipes, and an allowance of 3000 watts for the hot water supply.

boiler power = 4778.917 + 1194.729 + 3000

boiler power = 8973.646 watts

= 9 kW (approx.)

Convective heating using the environmental-temperature concept

The following steps are taken:

1. Calculate the heat loss through the various elements of the structure using the difference between the inside environmental temperature and the outside air temperature and sum these to produce ΣQf.
2. Calculate the area of the entire enclosure ΣA (this should include any partitions, floors, or ceilings gaining heat from or losing heat to the adjacent rooms).
3. Calculate $\Sigma Qf/\Sigma A$ and from the following equation find the difference between the inside air temperature and the outside environmental temperature:

$$t_{ai} - t_{ei} = \Sigma Qf/4.8\ \Sigma A$$

where t_{ai} = inside air temperature °C

t_{ei} = inside environmental temperature °C

ΣQf = rate of heat transfer through the building fabric W

ΣA = total area of room surfaces in m^2

4. Calculate the heat loss by ventilation using the formula given previously:

$$Q_v = 0.333\ N\ V\ (t_{ai} - t_{ao})$$

5. Add the results of steps (1) and (4) to give the total heat loss.

Example 7.12 *Using the bungalow given in Fig. 7.6 and the same U values as given in example 7.10, calculate the rate of heat loss and the surface area of the radiator required for the dining room when the internal environmental temperature is to be 21°C and convective heating is to be used.*

Table 7.12 Dining room

Type of structure	Dimensions (m)	Area (m^2)	U value (W/m^2 K)	Temperature difference (K)	Heat loss (W)
External wall	9 × 2.4 deduct 2 × 1.8 (window)	18.0	0.4	22	158.40
Window	2 × 1.8	3.6	2.8	22	221.76
Partition wall	3 × 2.4	7.2	2.2	3	47.52
Floor	3 × 3	9.0	0.2	22	39.6
Ceiling	3 × 3	9.0	0.3	22	59.40
	ΣA	46.8		ΣQf	526.68

internal air temperature

$$t_{ai} - t_{ei} = \frac{\Sigma Qf}{4.8\ \Sigma A}$$

$$t_{ai} - t_{ei} = \frac{526.68}{4.8 \times 46.8}$$

$$t_{ai} - t_{ei} = 2.34$$

$$t_{ai} = 21 + 2.34$$

$$= 23.3^\circ\text{C (approx.)}$$

heat loss by ventilation

$$Q_v = 0.333\ N\ V\ (t_{ai} - t_{ao})$$

$$= 0.333 \times 2 \times 21.6 \times 25$$

$$= 359.64$$

$$\text{total heat losses} = 526.68 + 359.64$$

$$= 886.32 \text{ watts}$$

$$\text{area of radiator} = \frac{886.32}{530}$$

$$= 1.7\ \text{m}^2 \text{ (approx.)}$$

Radiant heating using the environmental-temperature concept

In order to find the total heat losses the following steps are taken:

1. Calculate the heat loss through the structure as for convective heating to produce ΣQf.
2. Calculate the surface area of the entire enclosure as for convective heating to produce ΣA.
3. Find the ventilation conduction C_v from:

$$C_v = \frac{1}{0.333\ N\ V} + \frac{1}{4.8\ \Sigma A}$$

where N = number of air changes per hour

V = volume of the enclosure m^2

4. Calculate ventilation loss Q_v from:

$$Q_v = C_v\ (t_{ei} - t_{ao})$$

5. Add the results of steps 1 and 4 to give the total heat loss.

Example 7.13. *Using the bungalow given in Fig. 7.6 and the same U values given in example 7.10, calculate the rate of heat loss and the surface area of a radiant panel required for the dining room when the internal environmental temperature is to be 21 °C.*

$$\Sigma Qf = 667.80 \text{ W}$$

$$\Sigma A = 46.8\ \text{m}^2$$

$$N = 2 \text{ per hour}$$

$$V = 21.6\ \text{m}^3$$

ventilation conductance

$$\frac{1}{C_v} = \frac{1}{0.333\ N\ V} + \frac{1}{4.8\ \Sigma A}$$

$$= \frac{1}{0.333 \times 2 \times 21.6} + \frac{1}{4.8 \times 26.8}$$

$$= 0.07 + 0.008$$

$$C_v = 12.82 \text{ W K}$$

heat loss by ventilation

$$Q_v = C_v\ (t_{ei} - t_{ao})$$

$$= 12.82\ (21 - (-1))$$

$$= 12.82 \times 22$$

$$= 282.04 \text{ W}$$

total heat loss

$$Q_t = \Sigma Qf + \Sigma Q_v$$

$$= 564.3 + 282.04$$

$$= 746.34$$

$$\text{area of radiant panel} = \frac{746.34}{530}$$

$$= 1.4\ \text{m}^2 \text{ (approx.)}$$

Note: The heat required for convective heating is higher than the heat required for radiant heating. The human body loses heat by radiation, convection and evaporation by the following approximate percentages:

radiation	= 45 per cent
convection	= 30 per cent
evaporation	= 25 per cent

Since the greatest heat lost is by radiation, the provision of radiant heating will 'balance' this loss and will provide thermal comfort for the occupants of the room at a lower heat input.

Although the term radiant heating is used, convective heating is still provided by the radiant panels which should be sufficient to 'balance' the heat lost by convection.

The heat lost by evaporation can only be controlled by: (*a*) correct relative humidity of the air which is usually between 40 per cent and 60 per cent, and an air-conditioning system would be required for this purpose; (*b*) by the correct air velocity which is about 0.2 m/s maximum.

Heat emission from a radiator

Manufacturers usually give the radiator heat emission for a temperature differential between the mean water temperature and the ambient, or surrounding, air temperature of 55.6 °C.

For a temperature differential of other than 55.6 °C, the radiator heat emission may be found from the following formula:

$$E_2 = E_1 \times \left[\frac{\Delta t}{55.6}\right]^{1.3}$$

where E_2 = heat emission in W/m^2 for the new temperature differential

E_1 = heat emission in W/m^2 for a 55.6 °C differential

Δt = the new temperature differential in °C

Example 7.14. *Calculate the heat emission from a radiator in W/m^2 when the temperatures of the flow and return waters are 80 °C and 70 °C respectively, when the ambient air temperature is 18 °C. The heat emission for the radiator given by the manufacturers for a temperature differential of 55.6 °C is 520 W/m^2.*

$$\Delta t = \left[\frac{80 + 70}{2}\right] - 18$$

$$\Delta = 75 - 18 = 57\,°C$$

Substituting in the formula,

$$E_2 = 520 \times \left[\frac{57}{55.6}\right]^{1.3}$$

$$= 520 \times 1.025^{1.3}$$

$$= 520 \times 1.030$$

$$= 535.6\ \text{W/m}^2$$

Heat emission from a convector

When considering the heat emission from natural convectors for temperature differentials other than 55.6 °C, the following formula may be used:

$$E_2 = E_1 \times \left[\frac{\Delta t}{55.6}\right]^{1.5}$$

If the heat emitter in example 7.13 is a natural convector, the heat emission would be:

$$E_2 = 520 \left[\frac{57}{55.6}\right]^{1.5}$$

$$= 539.76\ \text{W/m}^2$$

Questions

1. Define the following terms: (*a*) thermal conductivity; (*b*) thermal resistivity; (*c*) thermal conductance; (*d*) thermal resistance; (*e*) thermal transmission.

2. One wall of a building consists of 25 mm thickness of cedar boardings; 76 mm thickness of glass wool; and 13 mm thickness of plasterboard. Using the following values, calculate the thermal transmission (U) for the wall.

Thermal conductivity	
Cedar wood	0.14 W/m K
Glass wool	0.042 W/m K
Plasterboard	0.58 W/m K
Thermal resistances	
Inside surface	0.123 m^2 K/W
Outside surface	0.055 m^2 K/W

Answer: 0.457 W/m^2 K

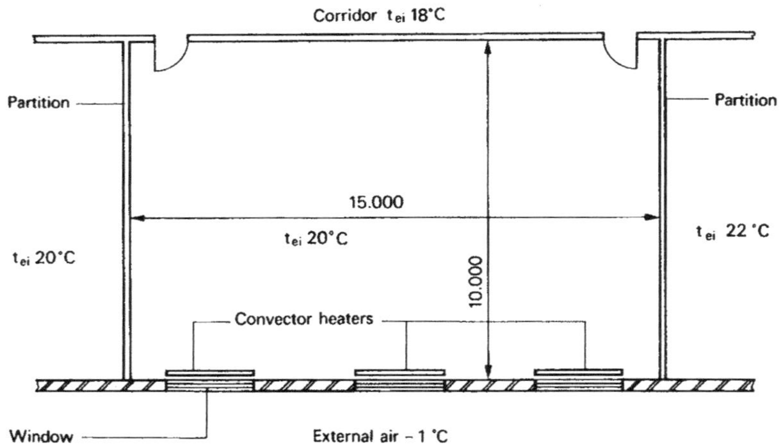

Fig.7.7 Plan of an office

3. If the thermal transmission for a wall is 0.74 W/m^2 K, calculate the heat loss in watts through the wall having an area of 36 m^2.

Answer: 2.664 watts

4. Figure 7.7 shows the plan of a single storey office having a floor to ceiling height of 3 m. The internal environmental temperature is 20 °C when the external air temperature is -1 °C.
Using the environmental-temperature concept for convective heating and and air change of 2 per hour, calculate the rating of the convector heaters in kW.

Use the following values:

(*a*) area of each window 4 m^2

(*b*) area of each door 1.4 m^2

(*c*) **U values (W/m^2 K)**

Floor	1.40
External walls	0.96
Partition walls	2.2
Windows	5.7
Doors	2.4
Ceiling	0.9

Answer: 17.4 kW (approx.). Use three heaters 5.8 kW each.

5. Using the following data, calculate the environmental temperature of a room in which the air temperature is 22 °C:

Fabric	Area (m^2)	Surface temperature (°C)
Floor	150	20
Walls	150	20
Windows	30	8
Ceiling	150	18

Answer: 19.73 °C

6. A pitched roof inclined at 30° has a horizontal ceiling below. If the U values for the roof and ceiling are found to be 2.80 W/m^2 K and 0.7 W/m^2 K respectively, calculate the combined U value for the ceiling and roof. Does this value satisfy the Building Regulations for offices?

Answers: 0.575 W/m^2 K; yes, it is below 0.6

7. Calculate the heat emission from a radiator in W/m^2 when the temperatures of the flow and return waters are 82 °C and 71 °C respectively, when the ambient air temperature is 16 °C. The heat emission for the radiator given by the manufacturers for a temperature differential of 55.6 °C is 530 W/m^2.

Answer: 591.5 W/m^2 (approx.)

8. Calculate the heat lost by ventilation in a room measuring 15 m by 10 m by 3 m, when the rate of air change is six per hour and the inside and outside air temperatures are 24 °C and −10 °C respectively.

Answer: 30.569 kW (approx.)

Chapter 8

Temperature drop through structures, condensation

Temperature drop through structures

To find the temperature distribution through a structure, it is necessary to know the following values:

1. the thermal resistances of the materials forming the structure;
2. temperature on either side of the structure.

Example 8.1 (see Fig. 8.1). *Calculate the U value and find the internal and external surface temperatures of a 105 mm thick solid brick wall when the internal and external air temperatures are 22 °C and 2 °C respectively.*

Use the following data:

external surface resistance = *0.053 m^2 K/W*

internal surface resistance = *0.123 m^2 K/W*

thermal conductivity of brick = *1.20 W/m K*

Thermal resistance of brick wall =

$$\text{Thermal resistance of brick wall} = \frac{L}{k} = \frac{0.105}{1.20} = 0.0875$$

$$U \text{ value} = \frac{1}{R_e + R_i + R_b}$$

$$= \frac{1}{0.053 + 0.123 + 0.0875}$$

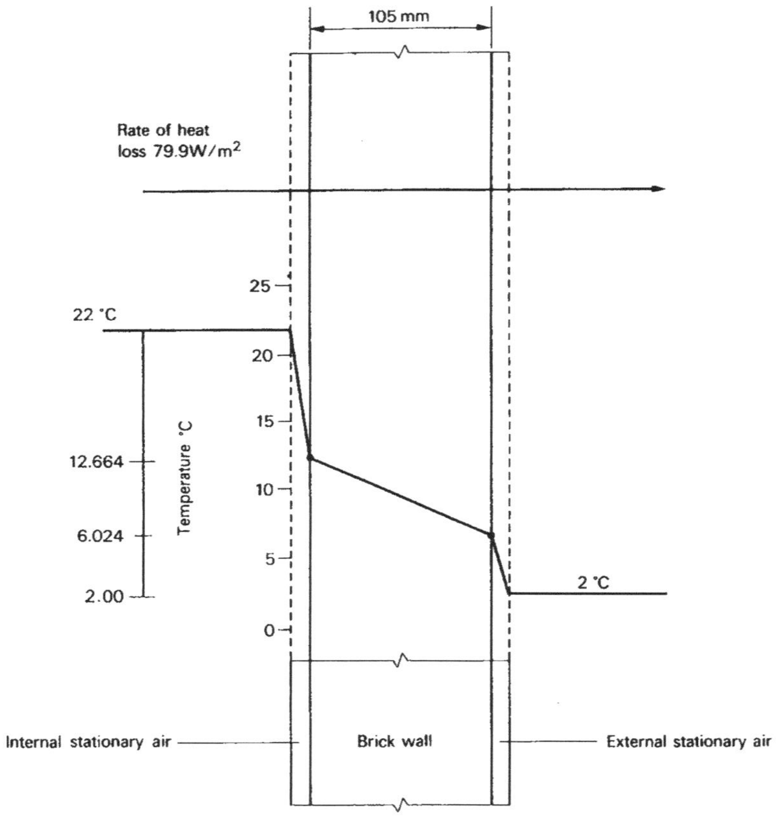

Fig.8.1 Graph of wall temperatures

$$= \frac{1}{0.2635}$$

$$= 3.796 \text{ W/m}^2\ K$$

rate of heat los $= U \times$ temperature difference

$$= 3.796 \times (22 - 2)$$

$$= 75.9 \text{ W/m}^2$$

If 75.9 W/m^2 is passing through the whole structure, it will pass through each separate resistance.

Internal surface of brick wall

rate of heat loss $= U \times$ temperature difference

$$= \frac{1}{R} \times \text{temperature difference}$$

$$75.9 = \frac{1}{0.123} \times \text{temperature difference}$$

∴ temperature difference $= 75.9 \times 0.123 = 9.336\,^\circ$C

temperature of stationary air or internal surface $= 22 - 9.336 = 12.664\,^\circ$C

External surface of brick wall

$$75.9 = \frac{1}{0.0875} \times \text{temperature difference}$$

∴ temperature difference $= 75.9 \times 0.0875 = 6.64\,^\circ$C

temperature of outer surface $= 12.664 - 6.64 = 6.024\,^\circ$C

External air (for checking purposes)

$$75.9 = \frac{1}{0.053} \times \text{temperature difference}$$

∴ temperature difference $= 75.9 \times 0.053 = 4.023\,^\circ$C

temperature of external air $= 6.024 - 4.023 = 2\,^\circ$C

Condensation

The terms used in problems of condensation are:

1. **Temporary condensation** This is condensation occurring on the internal surfaces when a sudden rise in the air temperature causes air in contact with surfaces to be temporarily at a much higher temperature than the surfaces. If the surfaces are below the dewpoint of the air, condensation will occur.
2. **Permanent condensation** In poorly insulated buildings the inside surfaces are at relatively low temperatures and if the internal air is comparatively humid, the internal surfaces may be at all times below the dewpoint temperature of the air. In these conditions condensation will be permanent.
3. **Interstitial condensation** This is condensation occurring within a structure. It is possible for interstitial condensation to occur although condensation on the internal surfaces is absent.

Example 8.2 (see Fig. 8.2). *If the brick wall in example 8.1 has 25 mm thickness of wood-wool slabs added to the internal surface, calculate the U value and find if interstitial condensation will occur if the dewpoint temperature of the air is 5 °C. The thermal conductivity of wood-wool is 0.09 W/m^2 K.*

$$\text{thermal resistance of wood-wool slabs} = \frac{0.025}{0.09} = 0.278$$

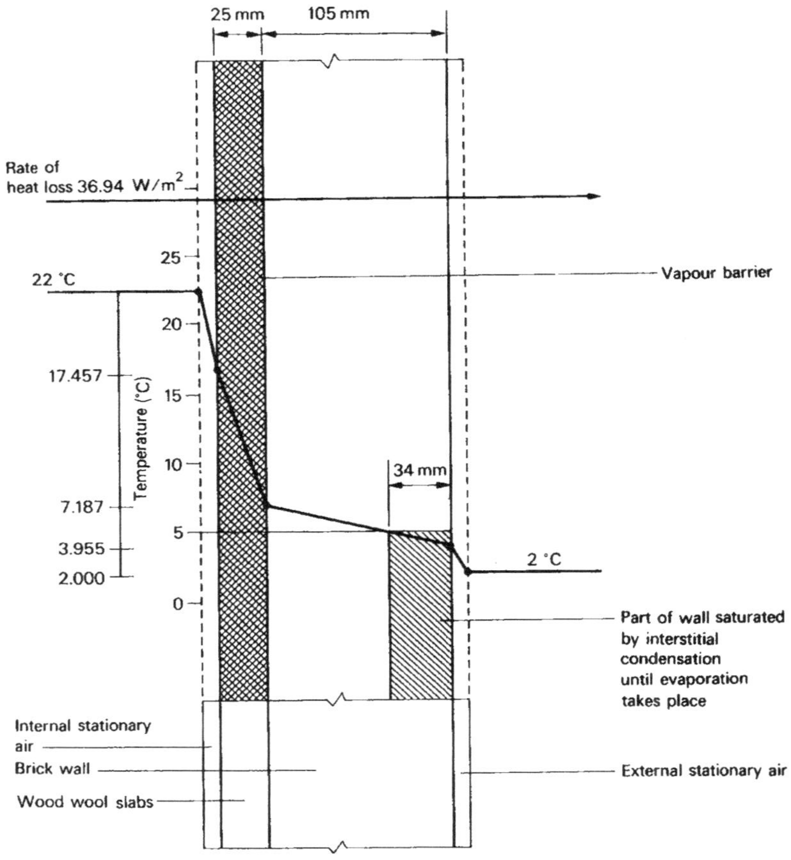

Fig. 8.2 Graph of wall temperatures and interstitial condensation

$$U \text{ value for wall} = \frac{1}{0.053 + 0.123 + 0.0875 + 0.278}$$
$$= \frac{1}{0.5415}$$
$$= 1.847 \text{ W/m}^2 \text{ K}$$

rate of heat loss for 1 m^2 of structure

$$= U \text{ value} \times \text{temperature difference}$$
$$= 1.847 \times (22 - 2)$$
$$= 36.94 \text{ watts}$$

Internal surface

$$\text{rate of heat loss} = U \times \text{temperature difference}$$
$$= \frac{1}{R} \times \text{temperature difference}$$
$$36.94 = \frac{1}{0.123} \times \text{temperature difference}$$
$$\text{temperature difference} = 36.94 \times 0.123 = 4.543\,^\circ\text{C}$$
$$\text{temperature of stationary air or internal surface} = 22 - 4.543 = 17.457\,^\circ\text{C}$$

Inner face of brick wall

$$36.94 = \frac{1}{0.278} \times \text{temperature difference}$$
$$\text{temperature difference} = 36.94 \times 0.278 = 10.27\,^\circ\text{C}$$
$$\text{temperature at inner face} = 17.457 - 10.27 = 7.187\,^\circ\text{C}$$

Outer face of brick wall

$$36.94 = \frac{1}{0.0875} \times \text{temperature difference}$$
$$\text{temperature difference} = 36.94 \times 0.0875 = 3.232\,^\circ\text{C}$$
$$\text{temperature at outer face} = 7.187 - 3.232 = 3.955\,^\circ\text{C}$$

Note: This is below dewpoint temperature and condensation will occur between 5 °C and 3.955 °C.

Temperature of external air

$$36.94 = \frac{1}{0.053} \times \text{temperature difference}$$
$$\text{temperature difference} = 36.94 \times 0.053 = 1.958\,^\circ\text{C}$$
$$\text{temperature of external air} = 3.955 - 1.958 = 1.997\,^\circ\text{C}$$

(The figure 1.997 °C instead of 2 °C is due to rounding off the previous values.)

Estimating condensation risk

In example 8.2 the dewpoint temperature was assumed, but The Building Research Establishment *Digest* 110 (Condensation) has made recommendations to form a basis for design which takes into account moisture content and ventilation rates.

At normal ventilation rates, the gain by the indoor air of body moisture from persons not engaged in physical exertion is roughly 4.5 g per person per hour. This results in the indoor air having an excess moisture content over outdoor air of 1.7 g of water vapour per kg of dry air.

Provided that ventilation rates are properly controlled, there would be a suitable design assumption for shops, offices, classrooms, assembly halls and dry industrial premises.

For dwellings, taking into account the moisture produced by cooking, etc., and the possible restricted ventilation in cold weather, a safer design value for excess moisture may be 3.4 g/kg.

Catering establishments and industrial workshops requiring humid atmospheres may contribute 6.8 g/kg or more to the inside air. In naturally ventilated premises, such design values may be added to assumed mixing ratio of the outdoor air.

The principles used in the *Digest* are to predict the likelihood of condensation, and to design so as to avoid this it may be applied to walls, floors, or roofs, but lightweight sheeted roofs present special problems.

Example 8.3 (see Fig. 8.3). *A cavity wall consists of 105 mm external brick leaf, 100 mm aerated concrete inner leaf, 16 mm of plaster and a 50 mm unventilated cavity. The internal and external air temperatures are 20 °C and 0 °C respectively. Calculate the U value and plot the structural temperature and dewpoint temperature through the wall. Check if interstitial condensation may take place. Use the following values:*

Thermal conductivities

Brick	*1.20 W/m K*
Aerated concrete	*0.14 W/m K*
Plaster	*0.40 W/m K*

Surface resistances

R_{si} *Internal surface layer* 0.123 m^2 K/W
R_{so} *External surface layer* 0.053 m^2 K/W
R_a *Air space* 0.18 m^2 K/W

Outside air saturated at a mixing ratio of 3.8 g/kg

1. The U value

$$U = \frac{1}{R_{si} + R_{so} + L_1/k_1 + L_2/k_2 + L_3/k_3 + R_a}$$

$$U = \frac{1}{0.123 + 0.053 + 0.105/1.20 + 0.100/0.140 + 0.016/0.40 + 0.18}$$

$$U = \frac{1}{0.123 + 0.053 + 0.087 + 0.714 + 0.04 + 0.18}$$

$$U = \frac{1}{1.1975}$$

$$= 0.835 \text{ W/m}^2 \text{ K}$$

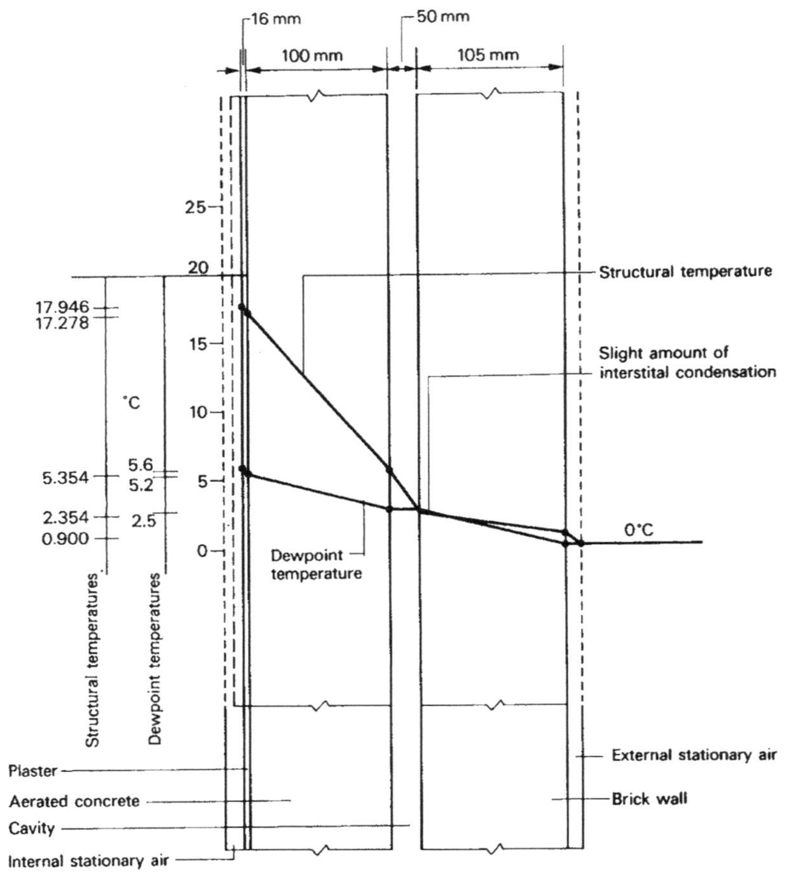

Fig.8.3 Graph of structural and dewpoint temperatures

2. Rate of heat loss through unit area

Heat loss = $U \times$ temperature difference
= 0.835 × 20
= 16.7 watts

3. Thermal resistances of structural components

(*a*) Internal stationary air	0.123 m^2 K/W
(*b*) Plaster	0.040 m^2 K/W
(*c*) Aerated concrete	0.714 m^2 K/W
(*d*) Brick	0.087 m^2 K/W
(*e*) External stationary air	0.053 m^2 K/W
(*f*) Air space	0.180 m^2 K/W

4. Calculation of temperatures

Since the heat loss of 16.7 watts must pass through each part of the cavity wall, taking each part separately, its thermal transmission (U) must be equal to 1/ resistance. The temperature at various points of the wall may therefore be found as follows:

$$\text{rate of heat loss} = U \times \text{temperature difference}$$

also $$\text{rate of heat loss} = \frac{1}{R} \times \text{temperature difference}$$

5. Internal stationary air or plaster surface

$$16.7 = \frac{1}{0.123} \times \text{temperature difference}$$

$$\begin{aligned} \text{temperature difference} &= 16.7 \times 0.123 \\ &= 2.054\,^\circ\text{C} \\ \text{temperature of plaster surface} &= 20 - 2.054 \\ &= 17.946\,^\circ\text{C} \end{aligned}$$

6. Inner face of aerated concrete leaf

$$16.7 = \frac{1}{0.04} \times \text{temperature difference}$$

$$\begin{aligned} \text{temperature difference} &= 16.7 \times 0.04 \\ &= 0.668\,^\circ\text{C} \\ \text{temperature at inner face} &= 17.946 - 0.668 \\ &= 17.278\,^\circ\text{C} \end{aligned}$$

7. Opposite face of aerated concrete leaf

$$16.7 = \frac{1}{0.714} \times \text{temperature difference}$$

$$\begin{aligned} \text{temperature difference} &= 16.7 \times 0.714 \\ &= 11.924\,^\circ\text{C} \\ \text{temperature at opposite face} &= 17.278 - 11.924 \\ &= 5.354\,^\circ\text{C} \end{aligned}$$

8. Inner face of brick leaf

$$16.7 = \frac{1}{0.18} \times \text{temperature difference}$$

$$\begin{aligned} \text{temperature difference} &= 16.7 \times 0.18 \\ &= 3\,^\circ\text{C} \\ \text{temperature at inner face} &= 5.354 - 3 \\ &= 2.354\,^\circ\text{C} \end{aligned}$$

9. Outer face of brick leaf

$$16.7 = \frac{1}{0.087} \times \text{temperature difference}$$

$$\begin{aligned} \text{temperature difference} &= 16.7 \times 0.087 \\ &= 1.453\,^\circ\text{C} \end{aligned}$$

$$\begin{aligned} \text{temperature at outer face} &= 2.354 - 1.453 \\ &= 0.9\,^\circ\text{C} \end{aligned}$$

10. External stationary air

$$16.7 = \frac{1}{0.053 \times \text{temperature difference}}$$

$$\begin{aligned} \text{temperature difference} &= 16.7 \times 0.053 \\ &= 0.885\,^\circ\text{C} \\ \text{temperature outside} &= 0.9 - 0.885 \\ &= 0.015\,^\circ\text{C} \end{aligned}$$

(The difference between the temperature given in the question as 0 °C is due to rounding off the figures.)

The section through the cavity wall can now be drawn to a suitable scale and the temperature difference across the structure set up on an adjacent vertical scale so that the various temperatures may be plotted.

Dewpoint temperature

A psychrometric chart is used to estimate dewpoints (see Fig. 8.4). The chart enables the effect of temperature changes and dewpoint temperatures across the wall to be predicted.

Condensation risk

At any point where the estimated structural temperature is lower than the dewpoint temperature, condensation may occur. In order to find the vapour-pressure drop and the corresponding dewpoint temperatures between points of the structure, the vapour resistivity of the various materials must be known. Table 8.1 gives both thermal and vapour resistivity of some common materials.

Table 8.1

Materials	**Thermal resistivity (m^2 K/W)**	**Vapour resistivity (MNs/g)**
Brickwork	0.7–1.4	25–100
Concrete	0.7	30–100
Rendering	0.8	100
Plaster	2	60
Timber	7	45–75
Plywood	7	1500–6000
Fibre building board	15–19	15–60
Hardwood	7	450–750
Plasterboard	6	45–60
Compressed strawboard	10–12	45–75
Wood-wool slabs	9	15–40
Expanded polystyrene	30	100–600
Foamed urea-formaldehyde	26	20–30
Foamed polyurethane (open or closed cell)	40–50	30–1000
Expanded ebonite	34	11 000–60 000

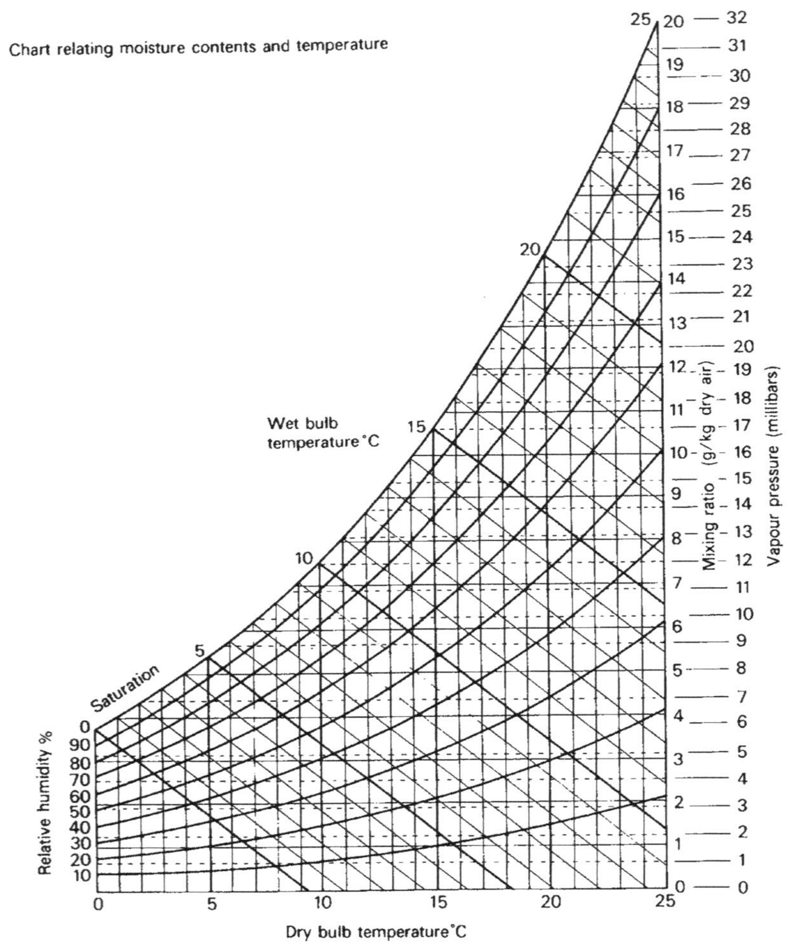

Fig.8.4 Psychrometric chart

1. From the psychrometric chart (Fig. 8.4) and using the design assumption that the external air is at 0 °C and is saturated at a mixing ratio of 3.8 g/kg, the outdoor vapour pressure can be read (on the right-hand scale) as 6 mb. If a moisture vapour excess of 1.7 g/kg is contributed to this by activities indoors, the two right-hand scales show that for a total moisture content of 3.8 + 1.7 = 5.5 g/kg and the indoor vapour pressure will be 9 mb. The difference between the indoor and outdoor vapour pressure is 9 − 6 = 3 mb, therefore ΔP = 3 mb.

2. Vapour resistance
(Value from Table 8.1 x thickness of component)

Plaster	60 x 0.016	= 0.96
Aerated concrete	33.45 x 0.100	= 3.45
Cavity	Nil	= 0.00
Brick	25 x 0.105	= 2.625
	R_v	= 6.930

3. Pressure drop between points

$$\frac{\Delta P}{R_v} \times r_v = \frac{3}{6.93} \times r_v = 0.433\, r_v \text{ (mb point to point)}$$

R_v = Total vapour resistance

r_v = Individual vapour resistances.

			Corresponding dewpoint temperature from chart (°C)
Indoor vapour pressure		9.000	= 5.6 °C
across plaster 0.433 x 0.96	=	0.416	
	drop to	8.584	= 5.2 °C
across inner leaf 0.433 x 3	=	1.299	
	drop to	7.285	= 2.5 °C
across cavity Nil	=	0.000	
		7.285	= 2.5 °C
across outer leaf 0.433 x 2.625	=	1.137	
	drop to	6.148	= 0.0 °C

Figure 8.3 shows a graph of these structural and dewpoint temperatures.

Note: If the outer portion of the wall is permeable to moisture, or if ventilation is provided behind permeable wall or roof claddings, condensation will not be troublesome because the moisture can evaporate gradually to the outside air. If the outer surface is impermeable, the condensed moisture tends to accumulate in the wall and may ultimately saturate the material. The situation will be most severe when the relative humidity inside is high.

Vapour barrier

A vapour barrier on the inner face of the wall (or the potentially warm side of any layer of insulating material) will prevent the passage of water vapour into the wall. If the outer face of the wall is more permeable than the inner vapour barrier, any moisture contained in the wall can evaporate to the outside air.

If, however, the outer face of the wall has an impermeable cladding, or if the cladding is of organic material (for example, timber) that would suffer in prolonged damp conditions, a ventilated cavity should be formed between the cladding and the wall so that any moisture evaporating from the wall surface is removed.

Surface temperature

The temperature at any point within a structure is proportional to the ratio which the thermal resistance at that point bears to the total thermal resistance of the structure.

If the U value is not required, the thermal resistance may be used to find the temperature drop through the structure:

$$\frac{R_s}{R_T} = \frac{t_s}{t_T}$$

where R_s = surface resistance

R_T = total thermal resistance of structure

t_s = temperature difference between the surface and the air

t_T = total temperature difference between inside and outside air

Example 8.4. *Find the internal temperature of 4 mm single glazing when the external temperature is 2 °C and the internal temperature is 20 °C.*

Use the following data:

external surface resistance	*0.053 m^2 K/W*
internal surface resistance	*0.123 m^2 K/W*
thermal conductivity of glass	*1.02 W/m K*

First find the total thermal resistance.

external surface resistance	0.053
internal surface resistance	0.123
thermal resistance of glass	
$= \frac{L}{k} = \frac{0.004}{1.02} =$	0.004
total thermal resistance	0.180

$$\frac{R_s}{R_T} = \frac{t_s}{t_T}$$

$$\frac{0.123}{0.180} = \frac{t_s}{(20-2)}$$

$$\therefore \quad 0.683 = \frac{t_s}{18}$$

$$t_s = 0.683 \times 18 = 12.294\,°C$$

internal surface temperature $= 20 - 12.3$
$= 7.7\,°C$ (approx.)

Example 8.5. *If in example 8.4 the wet-bulb temperature of the inside air is 14 °C, find by use of the psychrometric chart if condensation will occur on the inside surface of the glass.*

By intersecting the dry-bulb and wet-bulb temperature lines as shown in Fig. 8.4, it will be seen that the relative humidity of the inside air is about 50 per cent.

If a horizontal line is now drawn across from the point of intersection of the dry-bulb and wet-bulb temperature lines until it intersects the saturation curve, it will be seen that the temperature at which condensation will occur is about 9.8 °C. Condensation will therefore occur on the inside surface of the glass.

Example 8.6. *If the window in example 8.4 is double-glazed by adding another 4 mm thick pane of glass and leaving a 20 mm sealed air space between the two panes, find the surface temperature of the inner pane of glass.*

Use the same values given in example 8.4 but include:

thermal conductivity of air = 0.029 W/m °C

If the temperature of the inside air is to be 20 °C with a relative humidity of 50 per cent check by use of the psychrometric chart if condensation on the inside surface of the glass has now been prevented.

Total thermal resistance:

external surface resistance	= 0.053
internal surface resistance	= 0.123
thermal resistance of glass	
$= \frac{L}{k} = \frac{0.008}{1.02}$	= 0.008
thermal resistance of air space	
$= \frac{L}{k} = \frac{0.02}{0.029}$	= 0.689
total thermal resistance	= 0.873

$$\frac{R_s}{R_T} = \frac{t_s}{t_T}$$

$$\frac{0.123}{0.873} = \frac{t_s}{(20-2)}$$

$$t_s = 0.141 \times 18 = 2.538\,°C$$

internal surface temperature $= 20 - 2.538$
$= 17.5\,°C$ (approx.)

Since the air temperature is higher than the dewpoint temperature of 9.8 °C, condensation should not now occur.

Example 8.7. *A roof is to be constructed of 10 mm thick asbestos-cement sheets fixed to a steel frame and insulated with 50 mm thick fibre board to form an air space. If the inside and outside air temperatures are 20 °C and 0 °C respectively, calculate the U value and the inside surface temperature. Check by use of the psychrometric chart if condensation will occur on the inside surface of the insulation when the relative humidity of the inside air is 60 per cent.*

Use the following data:

thermal resistance of external air	*0.045 m^2 K/W*
thermal resistance of internal air	*0.110 m^2 K/W*

thermal resistance of air space	*0.18 m² K/W*
thermal resistivity of fibre board	*19 m K/W*
thermal resistivity of asbestos cement	*4.5 m K/W*

$$U = \frac{1}{R_{ao} + R_{ai} + R_a + (L_1 \times r_1) + (L_2 \times r_2)}$$

$$U = \frac{1}{0.045 + 0.110 + 0.18 + (0.01 \times 4.5) + (0.05 \times 19)}$$

$$U = \frac{1}{0.045 + 0.110 + 0.18 + 0.045 + 0.95}$$

$$U = \frac{1}{1.33}$$

$$= 0.752\ \text{W/m}^2\ \text{K}$$

$$\text{rate of heat loss} = 0.752 \times 20 = 15.04\ \text{W/m}^2$$

Temperature of inside surface

$$\text{rate of heat loss} = \frac{1}{R} \times \text{temperature difference}$$

$$\text{temperature difference} = 15.04 \times 0.110 = 1.654\,^\circ\text{C}$$

$$\text{temperature of inside surface} = 20 - 1.654 = 18.35\,^\circ\text{C}$$

Inspection of the psychrometric chart (Fig. 8.4) shows that for 20 °C indoor air temperature and 60 per cent relative humidity, the dewpoint temperature of the air is about 11.75 °C. The structural temperature of 18.35 °C is above the dewpoint temperature of 11.75 °C and therefore condensation on the surface is unlikely to occur. A vapour barrier behind the fibre-board insulation would, however, be a recommendation in case of interstitial condensation.

Alternatively, the temperature of the inside surface may be found from the thermal resistances as follows:

$$\frac{R_s}{R_T} = \frac{t_s}{t_T}$$

$$\frac{0.11}{1.33} = \frac{t_s}{20}$$

$$0.0827 = \frac{t_s}{20}$$

$$\therefore \quad t_s = 0.827 \times 20 = 1.654\,^\circ\text{C}$$

$$\text{temperature of inside surface} = 20 - 1.654 = 18.35\,^\circ\text{C}$$

Example 8.8. *A factory is to be constructed of corrugated sheeting fixed to a steel frame and insulated with fibre board lining to form an air space. It is required to provide sufficient thickness of insulation to prevent the inside* ***surface from falling below 18 °C when the inside and outside air temperatures are 20 °C and - 2 °C respectively.***

Using the following data, determine the minimum thickness of the insulation and the overall thermal transmittance.

U values

thermal transmittance of sheeting	*8 W/m² K*
thermal resistance of air space	*0.180 m² K/W*
thermal resistance of internal surface	*0.123 m² K/W*
thermal resistance of external surface	*0.053 m² K/W*
thermal resistivity of fibre board	*19 m K/W*

Thermal resistance of sheeting

$$\text{rate of heat loss} = U \times \text{temperature difference}$$

$$= 8 \times [20 - (-2)]$$

$$= 8 \times 22$$

$$= 176\ \text{W/m}^2$$

$$\text{rate of heat loss} = \frac{1}{R} \times \text{temperature difference}$$

$$176 = \frac{1}{R} \times 22$$

$$\frac{176}{22} = \frac{1}{R}$$

$$\therefore \quad R = \frac{22}{176}$$

$$= 0.125\ \text{m}^2\,^\circ\text{C/W}$$

thermal resistance of fibre board = resistivity × thickness in metres

$$\text{total resistance } R_T = 0.123 + 0.053 + 0.125 + 0.18 + (r \times \text{thickness})$$

$$= 0.481 + (r \times \text{thickness})$$

$$= 0.481 + (19 \times \text{thickness})$$

$$\frac{R_s}{R_T} = \frac{t_s}{t_T}$$

$$\frac{0.123}{0.481 + (19 \times \text{thickness})} = \frac{20 - 18}{20 - (-2)}$$

$$\frac{0.123}{0.481 + (19 \times \text{thickness})} = 0.09$$

$$0.09\,[0.481 + (19 \times \text{thickness})] = 0.123$$

$$(0.09 \times 0.481) + (0.09 \times 19 \times \text{thickness}) = 0.123$$

$$0.0433 + (1.71 \times \text{thickness}) = 0.123$$

$$1.71 \times \text{thickness} = 0.123 - 0.0433$$

$$\text{thickness} = \frac{0.123 - 0.0433}{1.71}$$

$$= 0.047\text{m}$$

$$= 50\text{mm (approx)}$$

$$U = \frac{1}{0.123 + 0.053 + 0.125 + 0.18 + (19 \times 0.05)}$$

$$U = \frac{1}{1.123 + 0.053 + 0.125 + 0.18 + 0.95}$$

$$U = 0.698 \text{ W/m}^2 \text{ K}$$

Questions

1. Find the internal and external surface temperatures of a 114 mm thick solid brick wall when the internal and external air temperatures are 20 °C and 2 °C respectively. Use the following data:

external surface resistance	0.053 m^2 K/W
internal surface resistance	0.123 m^2 K/W
thermal conductivity of brick	1.20 W/m K

Answers: internal surface of brick wall = 11.83 °C; external surface of brick wall = 5.52 °C

2. Define the following forms of condensation: (*a*) temporary; (*b*) permanent; (*c*) interstitial.

3. Find the internal and external surface temperatures of a cavity wall having a 105 mm thick brick outer leaf, 50 mm cavity and a 38 mm thick mineral wool slab inner leaf. Use the following values:

external surface resistance	0.053 m^2 K/W
internal surface resistance	0.123 m^2 K/W
thermal conductivity of brick	1.20 W/m K
thermal resistance of air space	0.18 m^2 K/W
thermal resistivity of mineral wool	8.66 m K/W

internal and external air temperatures 20 °C and 0 °C respectively

Answers: (*a*) internal surface = 16.8 °C
(*b*) opposite side of insulation = 8.27 °C
(*c*) inner surface of brick leaf = 3.62 °C
(*d*) outer surface of brick leat = 1.37 °C

4. Find the internal temperature of 8 mm thick single glazing when the internal and external air temperatures are 20 °C and −1 °C respectively. Use the following data:

external surface resistance	0.053 m^2 K/W
internal surface resistance	0.123 m^2 K/W
thermal conductivity of glass	1.20 W/m K

Answer: 5 7 °C (approx.)

5. If the window in question 4 is double-glazed by adding another 8 mm thick pane of glass and leaving a 20 mm sealed air space between the two panes, find the surface temperature of the inner pane of glass. Use the same values given in question 4 but including: thermal conductivity of air = 0.029 W/m K. If the room has a relative humidity of 50 per cent determine from the pschyrometric chart whether condensation will be deposited on the inside surface.

Answer: 17 °C. There will be no condensation as the dewpoint temperature is 9.5 °C.

6. A cavity wall consists of 105 mm thick brick outer leaf, 100 mm thick aerated concrete inner leaf, 16 mm internal plastering and a 50 mm cavity filled with foamed polyurethane. If the internal and external air temperatures are 22 °C and 0 °C respectively, find the structural and dewpoint temperatures through the wall and check the possibility of interstitial condensation taking place. Draw a section through the wall showing a graph of the temperatures. Use the following values:

thermal conductivities	
brick	1.20 W/m K
aerated concrete	0.14 W/m K
plaster	0.40 W/m K
thermal resistivity of foamed polyurethane	40 m K/W
internal surface resistance	0.123 m^2 K/W
external surface resistance	0.053 m^2 K/W
vapour resistances	
brick	60 MNs/g
polyurethane foam	40 MNs/g
aerated concrete	30 MNs/g
plaster	60 MNs/g

Outside air relative humidity 100 per cent. Inside activities assumed to contribute a moisture vapour excess of 3.2 g/kg.

Answer:

Position	**Structural temperature (°C)**	**Dewpoint temperature (°C)**
Surface of plaster	21.107	8.75
Inner face of aerated concrete leaf	20.817	8.00
Opposite face of aerated concrete leaf	15.633	6.20
Inner face of brick leaf	1.113	5.00
External face of brick leaf	0.483	0.00

Figure 8.5 shows a graph of structural and dewpoint temperatures.

7. A wall is to be constructed of 10 mm thick asbestos-cement sheets fixed to a steel frame and insulated with 50 mm thick fibre board lining to form an air space. If the inside and outside air temperatures are 20 °C and 0 °C respectively, calculate the inside surface temperature and check by use of the psychrometric chart if condensation will occur on the inside surface of the insulation when the relative humidity of the inside air is 50 per cent. Use the following data:

thermal resistance of external air	0.053 m^2 K/W
thermal resistance of internal air	0.123 m^2 K/W

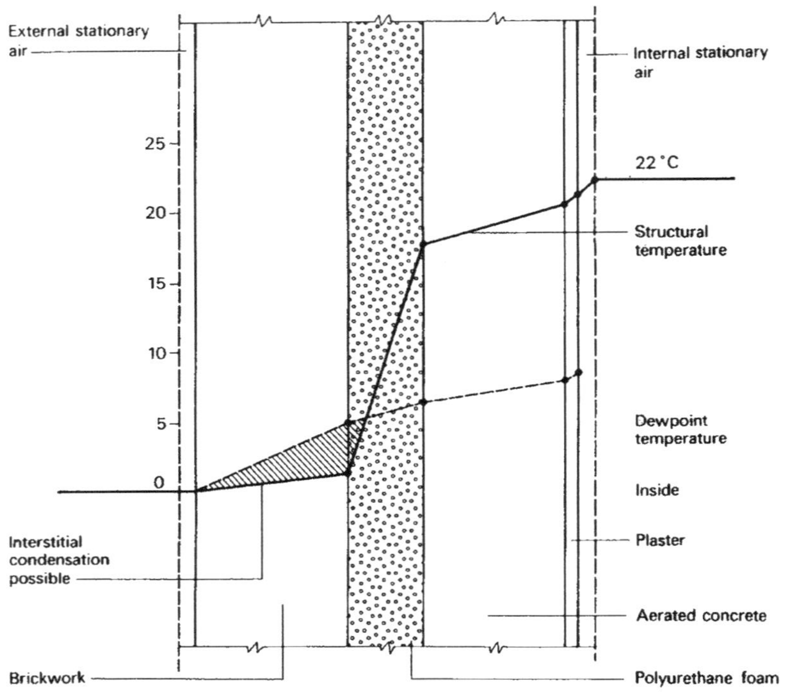

Fig.8.5 Graph of structural and dewpoint temperatures

thermal resistance of air space	0.180 m^2 K/W
thermal resistivity of asbestos cement	4.5 m K/W
thermal resistivity of fibre board	19 m K/W

Answers: Inside surface temperature = 18.18 °C
Dewpoint temperature of inside air = 9.8 °C
Condensation will therefore not occur

Chapter 9

Pipe sizing for heating, pump duty

Pumped circulation

Pumped circulation is used for all medium and large installations and has the following advantages over natural or gravity circulation:

1. Quicker response to heating of the building.
2. Circulation is independent of the temperature difference between the flow and return water.
3. Smaller pipes may be used, thus saving in capital costs, saving in space, and neater in appearance.
4. The boiler room may be sited on the roof.
5. Any type of heat emitter may be used.

Although the cost of the pump or pumps and the cost of power for running them must be considered, the use of pumped circulation adequately compensates by the saving in the overall capital and running cost of the installation.

The following steps may be taken in finding the sizes of the heating pipes:

1: Calculate the heat requirements for each room as described in Chapter 7.
2. Make a sketch of the system showing the positions of the heat emitters, boiler and the pipe runs. At this stage it will be necessary to decide on whether the system is to be one- or two-pipe and also the number of circuits to serve the various rooms in which heat emitters are to be installed.
3. Decide on the temperature of the flow and return water, i.e. the tempera-drop. Although this may appear arbitrary, the diameter of the pipe and the pump duty found later will provide for this temperature drop.

4. Estimate the heat emission to be expected from the pipes. It is usual to allow between 5 and 30 per cent of the total heat from the emitters for the heat emission from the pipes. The amount will depend upon the diameter and length of pipes.
5. Estimate the mass rates of flow through the circuit or circuits in kg/s.
6. Decide upon the velocity of water flow.
7. From the pipe-sizing table find a suitable size pipe or pipes.
8. Measure the actual lengths of the circuit or circuits and to this add an allowance for frictional losses due to fittings, to obtain the effective pipe length.
9. Find the pressure required for the pump.

An example will show how these steps are carried out.

Example 9.1. *Figure 9.1 shows a single-pipe circuit. Assuming each radiator has a heat emission of 8 kW and the temperature drop across the system is 15 °C, find by the use of Table 9.1 the approximate diameter of a steel pipe for the main and the pump pressure. Use a velocity of flow of water of 1 m/s.*

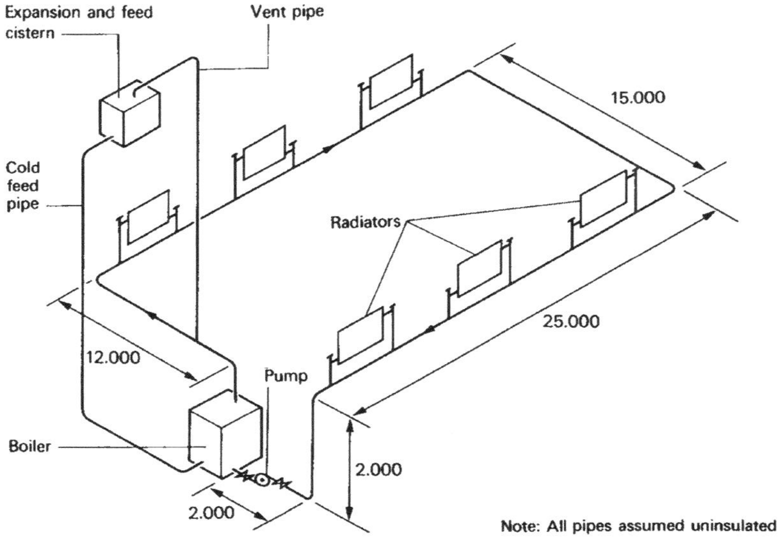

Fig.9.1 Single-pipe ring system

The single-pipe circuit must carry sufficient energy in the water flowing through it to supply the total heat required for the radiators, plus the heat required for the pipes.

total heat emission from radiators = 48 kW
heat emission from the pipes assuming 20 per cent of the heat emission from radiators = 9.6 kW
total = 57.6

Mass flow rate (kg/s)

$$\text{power} = \text{kg/s} \times \text{s.h.c.} \times (t_f - t_r) \quad \text{(kW)}$$

$$\therefore \text{flow rate} = \frac{\text{kW}}{\text{s.h.c.} \times (t_f - t_r)} \quad \text{(kg/s)}$$

where kW = total heat carried by the pipe
s.h.c. = specific heat capacity of water taken as 4.2 kJ/kg °C
t_f = temperature of flow water °C
t_r = temperature of return water °C

$$\text{flow rate} = \frac{57.6}{4.2 \times 15}$$

$$= 0.92 \text{ kg/s}$$

Table 9.1 Flow of water at 75°C in black steel pipes. Medium-grade pipe to BS 1387:1967

	20 mm		25 mm		32 mm		40 mm		50 mm			Pressure drop (Pa per m run of pipe)
V	M	EL	M	EL	M	EL	M	EL	M	EL	V	
	0.122	0.7	0.228	1.0	0.480	1.4	0.720	1.8	1.35	2.4		80.0
	0.124	0.7	0.232	1.0	0.488	1.4	0.732	1.8	1.37	2.4		82.5
	0.126	0.7	0.236	1.0	0.496	1.4	0.743	1.8	1.39	2.4		85.0
	0.128	0.7	0.240	1.0	0.503	1.4	0.755	1.8	1.41	2.4		87.5
	0.130	0.7	0.243	1.0	0.511	1.4	0.766	1.8	1.43	2.4		90.0
	0.132	0.7	0.247	1.0	0.518	1.5	0.778	1.8	1.45	2.4		92.5
	0.134	0.7	0.251	1.0	0.526	1.5	0.789	1.8	1.48	2.4		95.0
	0.136	0.7	0.254	1.0	0.533	1.5	0.800	1.8	1.50	2.4		97.5
	0.138	0.7	0.258	1.0	0.540	1.5	0.810	1.8	1.52	2.4		100.0
	0.152	0.7	0.284	1.0	0.595	1.5	0.893	1.8	1.67	2.4		120.0
	0.165	0.8	0.308	1.0	0.646	1.5	0.968	1.8	1.81	2.5		140.0
0.5	0.178	0.8	0.331	1.0	0.693	1.5	1.04	1.8	1.94	2.5		160.0
	0.189	0.8	0.353	1.0	0.738	1.5	1.11	1.8	2.06	2.5	1.0	180.0
	0.200	0.8	0.373	1.0	0.780	1.5	1.17	1.9	2.18	2.5		200.0
	0.211	0.8	0.392	1.1	0.820	1.5	1.28	1.9	2.29	2.5		220.0
	0.221	0.8	0.411	1.1	0.858	1.5	1.29	1.9	2.40	2.5		240.0
	0.230	0.8	0.428	1.1	0.895	1.5	1.34	1.9	2.50	2.5		260.0
	0.239	0.8	0.445	1.1	0.931	1.5	1.39	1.9	2.60	2.6		280.0
	0.248	0.8	0.462	1.1	0.965	1.5	1.44	1.9	2.69	2.6		300.0
1.0	0.257	0.8	0.478	1.1	0.998	1.6	1.49	1.9	2.78	2.6	1.5	320.0

V = velocity of flow m/s
M = mass rate of flow kg/s
EL = equivalent length factor

From the pipe-sizing table (Table 9.1), a 40 mm diameter steel pipe will carry 0.968 kg/s at a velocity of flow of 1 m/s giving a pressure loss per metre run of 140 Pa and an EL factor of 1.8.

Check on the estimated heat loss on the pipes of 20 per cent

Table 9.2 shows that for a 50 °C difference between the pipe surface and the surrounding air, the heat loss on a 40 mm diameter pipe is 105 watts per metre run.

Table 9.2 Theoretical heat emission from single horizontal steel pipes freely exposed in ambient air

Nominal bore (mm)	Temperature difference of surface to surroundings (°C)				
	40	45	50	55	60
8	28	32	35	39	43
10	33	37	41	46	50
15	40	46	53	59	66
20	48	56	62	70	78
25	58	68	78	88	98
32	71	82	93	110	120
40	78	92	105	110	130
50	96	112	130	150	170
65	120	140	160	180	200
80	140	160	180	211	230

Heat emission (W/m run)

Since the total length of pipe is 82 m,

$$\text{heat emission from pipe} = 82 \times 105 = 8610 \text{ watts}$$

$$\text{percentage emission} = \frac{8610}{48\,000} \times \frac{100}{1} = 18 \text{ per cent}$$

This is only a 2 per cent difference between the estimated heat loss from the pipes and the actual heat loss, and therefore the pipe diameter is suitable.

Pump duty

Before the pressure required by the pump can be determined, it will be necessary to find the effective length of the pipe. The effective length is the addition of the equivalent length due to the resistances of fittings and the actual length.

The resistances of the fittings are computed, in terms of equivalent length of straight pipe of corresponding diameter, by the product of an equivalent length factor found from Table 9.1 and a velocity pressure-loss factor, *K*.

Table 9.3 gives the *K* factor for some of the common types of fittings.

The pressure created by a pump on a one-pipe system does not force water through the radiators (unless special injector tees are used). The hot water therefore circulates through the radiators by natural convection created by the difference in density of the water flowing into the radiators and the cooler, denser water flowing out.

The frictional resistances through the tees for the purpose of determining the effective length of pipe will therefore be ignored.

Equivalent length of pipe = number of fittings × *K* × EL.

From Table 9.3, the *K* factor for a boiler is 5; and for 40 mm diameter 90° bends, it is 0.5. The equivalent lengths are therefore:

Table 9.3 Values of velocity pressure factor *K* for pipe fittings and equipment

Tees	*K* factor
	0.5 plus 0.2 for enlargement or reduction
Gate valve	0.2
Angle valve	0.5
Radiator	5.0
Sectional boiler	5.0
90° elbow 10 mm to 25 mm	0.8
90° elbow 32 mm to 50 mm	0.7
90° elbow 65 mm to 90 mm	0.6
90° bend 10 mm to 25 mm	0.7
90° bend 32mm to 50 mm	0.5
90° bend 65 mm to 90 mm	0.4
Reductions or enlargements on a straight run of pipe	
3 : 2 reduction	0.3
2 : 1 reduction	0.4
3 : 1 reduction	0.4
4 : 1 reduction	0.5
3 : 2 enlargement	0.4
2 : 1 enlargement	0.7
3 : 1 enlargement	0.9

$$\text{boiler} = 1 \times 5 \times 1.8 = 9.0 \text{ m}$$
$$\text{bends} = 6 \times 0.5 \times 1.8 = 5.4 \text{ m}$$
$$14.4 \text{ m}$$

$$\text{actual length} = 1 + 12 + 25 + 15 + 25 + 2 + 2 = 82 \text{ m}$$
$$\therefore \text{ effective length of pipe} = 82 + 14.4 = 96.4 \text{ m}$$

Pump pressure

The pressure loss through the pipe from Table 9.1 was found to be 140 Pa per metre run of pipe.

$$\therefore \text{ total pressure loss} = 96.4 \times 140 = 13\,496 \text{ Pa}$$

This is approximately equal to 13.5 kPa pressure and is equal to $13.5 \div 9.81 = 1.4$ m head of water.

Pumps are ordered giving the pressure developed and the flow rate in kg/s or litre/s. A pump providing a pressure of 13.5 kPa and giving a mass flow rate of 1 kg/s would be satisfactory.

Two-pipe system

In a two-pipe system, the flow pipes will reduce in diameter as they pass along the circuit, and the return pipes will increase in diameter as they return to the boiler.

The pressure created by a pump in a two-pipe system acts through the heat emitters and therefore convector heaters, which offer more resistance to the flow of water than radiators, can be used. The frictional resistances of the tees will, however, have to be taken into account when determining the effective length of pipe and the pump pressure.

Index circuit

In order to determine the pump pressure, the index circuit has to be selected. This is the circuit having the greatest resistance to the flow of water and supplies the index heat emitter. For a pumped system the index circuit is always taken to be the longest circuit.

Heat distribution

Unlike the one-pipe system, where the cooler water from the heat emitter passes back into the flow pipe, in the two-pipe system, the cooler water from each heat emitter passes into the return pipe and this provides a better balance of heat distribution in the system.

Example 9.2. *Figure 9.2 shows a two-pipe parallel system to be used as an alternative to the one-pipe ring system given in example 9.1. Using the follow-data and assumptions find by use of Tables 9.1–9.3 the approximate diameters of the flow and return mains and pump duty.*

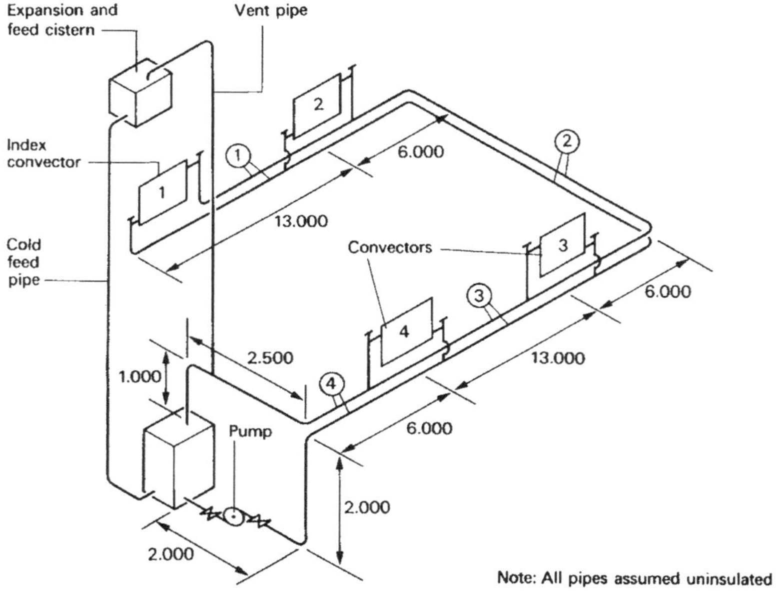

Fig.9.2 Two-pipe parallel system

1. *Heat emission from each convector* = *12 kW*
2. *Temperature drop across the system* = *15 °C*
3. *Mean temperature of water* = *72 °C*
4. *Specific heat capacity of water* = *4.2 kJ/kg °C*
5. *Velocity of flow of water* = *1 m/s*

Pipe sizing

Before the pipe-sizing procedure is commenced, a reference number or letter should be allocated to each section of pipework for identification purposes. The pipe-sizing may be started from the boiler or the farthest convector from it. In this example, the index convector is used as the starting point.

In order to find the mass flow rate through each section of pipework, an estimate is made of the carrying capacity of the pipes for each section, by taking the heat required for the convector and adding an estimated percentage for the pipework heat emission. This estimate can be checked and the pipe-sizing revised, if required.

Pipes number 1: Heat emission from convector 1 and pipework 1

emission from convector = 12.0 kW
emission from pipework 5 per cent of 48 kW = 2.4 kW
total = 14.4 kW

$$\text{flow rate} = \frac{14.4}{4.2 \times 15} = 0.228 \text{ kg/s}$$

From pipe-sizing Table 9.1, a 20 mm diameter steel pipe will carry 0.257 kg/s giving a pressure loss per metre run of 320 Pa and an EL factor of 0.8.

Check on estimated heat loss of 5 per cent assuming a temperature of pipe surface to the ambient air of 50 °C (see Table 9.2).

$$26 \text{ m} \times 62 = 1612 \text{ watts}$$

$$\text{percentage emission} = \frac{1612}{48\,000} \times \frac{100}{1}$$

$$\text{heat loss} = 3.4 \text{ per cent}$$

Since 5 per cent was allowed, pipes number 1 are satisfactory.

Pipes number 2: Emission from convector 2 and pipework 2

emission from convector = 12.0 kW
emission from pipework 10 per cent of 48 kW = 4.8 kW
total = 16.8 kW

$$\text{flow rate} = \frac{16.8 \text{ kW}}{4.2 \times 15} = 0{:}267 \text{ kg/s}$$

Pipes number 2 will also have to carry the heat required for emitter and pipework 1.

$$\text{total flow rate} = 0.228 + 0.267$$
$$= 0.5 \text{ kg/s}$$

From the pipe-sizing Table 9.1, a 32 mm diameter pipe will carry 0.503 kg/s giving a pressure loss per metre run of 87.5 Pa and an EL factor of 1.4.

Check on estimated heat loss of 10 per cent (see Table 9.2).

$$54 \text{ m} \times 93 = 5022 \text{ watts}$$

$$\text{percentage heat loss} = \frac{5022}{48\,000} \times \frac{100}{1}$$

$$\text{heat loss} = 10.46 \text{ per cent}$$

Pipes number 2 are therefore satisfactory for practical purposes.

Pipes number 3: Emission from convector and pipework 3

emission from convector = 12.0 kW
emission from pipework = 2.4 kW
5 per cent of 48 kW
total = 14.4 kW

$$\text{flow rate} = \frac{14.4}{4.2 \times 15}$$
$$= 0.228 \text{ kg/s}$$

Pipes number 3 will also have to carry the heat required for emitters and pipework 1 and 2.

$$\text{total flow rate} = 0.5 + 0.228$$
$$= 0.728 \text{ kg/s}$$

From pipe-sizing Table 9.1, a 32 mm diameter pipe will carry 0.738 kg/s giving a pressure loss per metre run of 180 Pa and an EL factor of 1.5.

Check on estimated heat loss of 5 per cent (see Table 9.2).

$$26 \text{ m} \times 93 = 2418 \text{ watts}$$

$$\text{percentage heat loss} = \frac{2418}{48\,000} \times \frac{100}{1}$$

$$\text{heat loss} = 5 \text{ per cent}$$

Pipes number 3 are therefore satisfactory.

Pipes number 4: Emission from convector 4 and pipework 4

emission from convector = 12.0 kW
emission from pipework = 2.4 kW
5 per cent of 48 kW
total = 14.4 kW

$$\text{flow rate} = \frac{14.4}{4.2 \times 15}$$
$$= 0.228 \text{ kg/s}$$

Pipes number 4 will also have to carry the heat required for emitters and pipework 1, 2 and 3.

$$\text{total flow rate} = 0.728 + 0.228$$
$$= 1 \text{ kg/s}$$

From pipe-sizing Table 9.1, a 40 mm diameter steel pipe will carry 1.04 kg/s giving a pressure loss per metre run of 160 Pa and an EL factor of 1.8.

Check on estimated heat loss of 5 per cent (see Table 9.2).

$$19.5 \text{ m} \times 105 = 2047.5 \text{ watts}$$

$$\text{percentage heat loss} = \frac{2047.5}{48\,000} \times \frac{100}{1}$$

$$\text{heat loss} = 4.3 \text{ per cent (approx.)}$$

Pipes number 4 are therefore satosfactory for practical purposes;

Pump duty

Before the pressure required by the pump can be determined, it is necessary to find the effective length of each section of pipework in the system. The effective lengths of pipe are given by addition of the equivalent length due to the resistances of the fittings and the actual lengths.

Effective lengths (see Table 9.3)

Pipes number 1, 20 mm diameter with an actual length of 26 m.

bends	= 2 × 0.7 × 0.8 = 1.12 m
angle valves	= 2 × 0.5 × 0.8 = 0.80 m
tees	= 2 × 0.7 × 0.8 = 1.12 m
emitter	= 1 × 5.0 × 0.8 = 4.00 m
total	= 7.04 m

effective length = 26 + 7.04 = 33.04 m

Pipes number 2, 32 mm diameter with an actual length of 54 m.

bends	= 2 × 0.5 × 1.4 = 1.40 m
angle valves	= 2 × 0.5 × 1.4 = 1.40 m
tees	= 2 × 0.7 × 1.4 = 1.96 m
emitter	= 1 × 5.0 × 1.4 = 7.00 m
total	= 11.76 m

effective length 54 + 11.76 = 65.76 m

Pipes number 3, 32 mm diameter with an actual length of 26 m.

tees	= 2 × 0.7 × 1.5 = 2.1 m
angle valves	= 2 × 0.5 × 1.5 = 1.5 m
emitter	= 1 × 5.0 × 1.5 = 7.5 m
total	= 11.1 m

effective length = 26 + 11.1 = 37.1 m

Pipes number 4, 40 mm diameter with an actual length of 18.5 m.

bends	= 3 × 0.5 × 1.8 = 2.70 m
angle valves	= 2 × 0.5 × 1.8 = 1.80 m
tees	= 2 × 0.7 × 1.8 = 2.52 m
emitter	= 1 × 5.0 × 1.8 = 9.00 m

boiler $= 1 \times 5.0 \times 1.8 = 9.00$ m

total $= 25.02$ m

effective length $= 18.5 + 25 = 43.50$ m

Pump pressure

The pressure developed by the pump will have to overcome the frictional resistances through all the effective lengths of pipes to the index convector. A table may now be compiled (see Table 9.4).

Table 9.4 Total pressure loss on the system

Pipe number	Effective pipe length (m)	Pressure loss per metre run (Pa)	Pressure loss on section (Pa)
1	33.04	320.0	10 572.8
2	65.76	87.5	575.4
3	11.10	180.0	1 998.0
4	25.02	160.0	4 003.2
		Total pressure loss	17 149.4 Pa

A pump providing a pressure of 17.2 kPa (approx.) and giving a mass flow rate of 1 kg/s would be satisfactory.

Boiler power

The boiler will have to have sufficient power to supply heat required from the pipes and the convectors.

$$\text{boiler power} = 14.4 + 16.8 + 14.4 + 14.4 = 60 \text{ kW}$$

A 10 per cent boiler margin may be added for pre-heating and to give a reserve of power for severe weather conditions.

Therefore the boiler power would be 66 kW.

Micro-pipe systems

Micro-pipe systems can be 'open' or 'closed' circuits. An open circuit is provided with an expansion and feed cistern with a cold feed and vent pipe, which are open to the atmosphere. A closed or sealed system dispenses with an expansion and feed cistern, and the open cold feed and vent pipes, and an expansion vessel is substituted in their place. This expansion vessel contains a diaphragm with nitrogen or air on one side and water from the heating system on the other. When the water is heated, the expansion is taken up by compressing the nitrogen or air. The Heating and Ventilating Contractors Association in its 'Guide to Good Practice' limits the flow-water temperature in open-circuit micro-pipe systems to 82 °C and in closed or sealed systems to 99 °C. Open circuits are usually operated with a temperature difference between the flow and return water of 11 °C, while sealed systems usually operate at a temperature difference between flow and return water of up to 20 °C. This higher temperature in the sealed systems increases the carrying capacity of the pipes.

The outside diameter of the pipes to the heat emitters may be 6 mm, 8 mm, or 10 mm, depending upon the heating load and the pump pressure.

Circuit design

The following factors are usually taken into account:

1. To prevent air-locking, the velocity of flow of water should be as high as possible; up to 1.5 m/s and above 0.3 m/s.
2. In determining the mass flow rate, an allowance is made for the heat loss from the pipework. This loss is generally taken as 5 per cent for the larger-diameter mains and 10 per cent for the micro-pipe, of the heat emission from the radiators each serve.

The following steps may be taken in finding the sizes of the heating pipes:

1. Calculate the heat requirements for each room as described in Chapter 7.
2. Make a sketch of the system showing the positions of the heat emitters, boiler and pipe runs. The micro-pipe system is always a two-pipe system.
3. Allocate a reference number to each section of pipework.
4. Decide on the temperature difference between the water in the flow and return pipes.
5. Estimate the mass flow rate through each circuit, allowing 5 or 10 per cent heat losses on the pipes.
6. Allowing a velocity of flow of water above 0.3 m/s and up to 1.5 m/s, find a suitable diameter pipe from Tables 9.5 and 9.6.
7. Measure the actual lengths of the circuits and to these add an allowance for frictional losses due to fittings to obtain the effective pipe length.
8. Find the 'index run' and determine the pressure drop along it in order to ascertain the pump pressure.

An example will show how these steps are carried out.

Example 9.3. *Figure 9.3 shows a micro-pipe system for a three-bedroomed house. If an 'open' type of system is to be used with flow and return water temperatures of 80 °C and 69 °C respectively, find the approximate diameters of the pipes and the pump duty.*

Pipes number 1

emission from radiator $= 0.900$ kW
emission from pipework $= 0.090$ kW
(10 per cent of 0.9 kW)

total $= 0.990$ kW

$$\text{flow rate} = \frac{0.99}{4.2 \times (80-69)} = 0.0214 \text{ kg/s}$$

From the pipe-sizing table (Table 9.5), an 8 mm outside diameter tube will carry 0.0217 kg/s giving a pressure loss per metre run of 902 Pa.

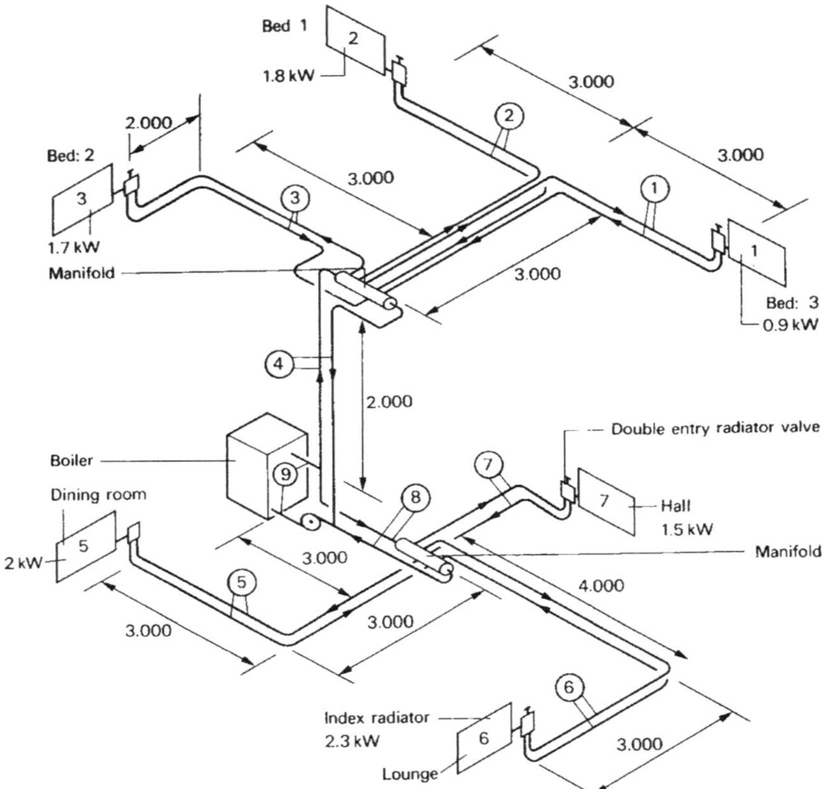

Note: All pipes assumed uninsulated : hot water cylinder omitted

Fig.9.3 Micro-pipe system

Pipes number 2

emission from radiator = 1.800 kW
emission from pipework = 0.180 kW
(10 per cent of 1.8 kW)

total = 1.980 kW

$$\text{flow rate} = \frac{1.98}{4.2 \times 11}$$

= 0.043 kg/s

From the pipe-sizing Table 9.5, a 10 mm outside diameter tube will carry 0.0433 kg/s giving a pressure loss per metre run of 902 Pa.

Pipes number 3

emission from radiator = 1.700 kW

Table 9.5 Table for the flow of water at 82 °C in small-diameter copper pipes (outside diameter)

Diameter 6 mm	Diameter 8 mm	Diameter 10 mm	Velocity	Pressure drop (Pa per m run of pipe)
(kg/s)	(kg/s)	(kg/s)	(m/s)	
0.005 6	0.013 8	0.027 7		410.0
0.006 3	0.015 3	0.030 6		492.0
0.006 8	0.016 8	0.033 5		574.0
0.007 4	0.018 2	0.036 2	0.6	656.0
0.007.9	0.019 5	0.038 8		738.0
0.008 4	0.020 6	0.041 0		820.0
0.008 8	0.021 7	0.043 3		902.0
0.009 3	0.022 9	0.045 6		984.0
0.009 6	0.024 0	0.047 6		1066.0
0.010 2	0.025 1	0.049 8	0.9	1148.0
0.010 5	0.026 1	0.051 6		1230.0
0.010 9	0.027 1	0.053 5		1312.0
0.011 3	0.027 9	0.055 4		1394.0
0.011 6	0.029 0	0.057 4		1476.0
0.012 1	0.029 8	0.058 7		1558.0
0.012 4	0.030 6	0.060 8		1640.0
0.013 3	0.032 8	0.065 1		1845.0
0.014 2	0.034 9	0.069 0		2050.0
0.015 0	0.036 8	0.072 9	1.2	2255.0

emission from pipework = 0.170 kW
(10 per cent of 1.7 kW)

total = 1.870 kW

$$\text{flow rate} = \frac{1.870}{4.2 \times 11}$$

= 0.040 kg/s

From the pipe-sizing Table 9.5, a 10 mm outside diameter tube will carry 0.041 kg/s giving a pressure loss per metre run of 820 Pa.

Pipes number 4

These pipes will have to carry the heat loads required for pipes 1, 2 and 3 plus the heat emission from the pipes number 4.

heat load pipes 1 = 0.990 kW
heat load pipes 2 = 1.980 kW
heat load pipes 3 = 1.870 kW

= 4.840 kW

emission from pipes number 4 = 0.242
(5 per cent of 4.84 kW)

total = 5.082 kW

$$\text{flow rate} = \frac{5.082}{4.2 \times 11}$$

= 0.110 kg/s

From pipe-sizing Table 9.6, a 22 mm outside diameter tube will carry 0.111 kg/s giving a pressure loss per metre run of pipe of 80 Pa.

Pipes number 5

emission from radiator = 2.00 kW
emission from pipes = 0.20 kW
(10 per cent of 2 kW)

total = 2.20 kW

$$\text{flow rate} = \frac{22}{4.2 \times 11}$$

= 0.048 kg/s

From the pipe-sizing Table 9.5, a 10 mm outside diameter tube will carry 0.0498 kg/s giving a pressure loss per metre run of 1148 Pa.

Pipes number 6

emission from radiator = 2.300 kW
emission from pipes = 0.230 kW
(10 per cent of 2.3 kW)

total = 2.530 kW

$$\text{flow rate} = \frac{2.530}{4.2 \times 11}$$

= 0.054 kg/s

From the pipe-sizing Table 9.5, a 10 mm outside diameter tube will carry 0.0554 kg/s giving a pressure loss per metre run of 1394 Pa.

Pipes number 7

emission from radiator = 1.500 kW
emission from pipes = 0.150 kW
(10 per cent of 1.5 kW)

total = 1.650 kW

$$\text{flow rate} = \frac{1.650}{4.2 \times 11}$$

= 0.036 kg/s

From the pipe-sizing Table 9.5, a 10 mm outside diameter tube will carry 0.0362 kg/s giving a pressure loss per metre run of 656 Pa.

Pipes number 8

These will have to carry the heat loads for pipes 5, 6 and 7 plus the heat emission from pipes number 8.

heat load pipes 5 = 2.200 kW
heat load pipes 6 = 2.530 kW
heat load pipes 7 = 1.650 kW

total = 6.380 kW

Pipes number 8 are very much shorter in length than pipes 5, 6 and 7, therefore the estimated heat emission from them will be taken as 5 per cent of the above total.

Heat load to be carried by pipes number 8 is therefore:

6.38 kW + 5 per cent of 6.38

= 6.38 + 0.319
= 6.699 kW

$$\text{flow rate} = \frac{6.699}{4.2 \times 11}$$

= 0.145 kg/s

From the pipe-sizing Table 9.6, a 22 mm outside diameter tube will carry 0.152 kg/s giving a pressure loss per metre run of 140 Pa.

Table 9.6 Table for the flow of water at 80 °C in copper pipes (outside diameter). Block table to BS 2871-1-X

Diameter 12 mm	**Diameter 15 mm**	**Diameter 22 mm**	**Diameter 28 mm**	**Diameter 35 mm**	**Velocity**	**Pressure drop (Pa per m run of pipe)**
(kg/s)	(kg/s)	(kg/s)	(kg/s)	(kg/s)	(m/s)	
0.016	0.030	0.089	0.181	0.181	0.3	55.0
0.017	0.032	0.094	0.191	0.344		60.0
0.017	0.033	0.098	0.199	0.360		65.0
0.018	0.035	0.103	0.208	0.376		70.0
0.019	0.036	0.107	0.216	0.390		75.0
0.020	0.037	0.111	0.224	0.405		80.0
0.021	0.040	0.119	0.240	0.433	0.5	90.0
0.022	0.043	0.126	0.254	0.459		100.0
0.025	0.047	1.140	0.282	0.508		120.0
0.027	0.052	0.152	0.308	0.554		140.0
0.030	0.056	0.164	0.332	0.597		160.0
0.032	0.060	0.176	0.354	0.638		180.0
0.034	0.064	0.186	0.376	0.676		200.0
0.038	0.071	0.207	0.416	0.749	1.0	240.0

Pipes number 9

These will have to carry the heat loads for pipes 1, 2, 3, 5, 6, and 7 plus the heat emission from pipes number 4 and 8.

heat load pipes 1 = 0.990 kW
heat load pipes 2 = 1.980 kW

heat load pipes 3 = 1.870 kW
heat load pipes 4 = 0.242 kW
heat load pipes 5 = 2.200 kW
heat load pipes 6 = 2.530 kW
heat load pipes 7 = 1.650 kW
heat load pipes 8 = 0.319 kW
11.781 kW

Pipes number 9 are again very short and an estimated heat loss from them of 5 per cent of the above total will be adequate.

heat load to be carried by pipes number 9 = 11.781 + 5 per cent of 11.781
= 11.781 + 0.589
= 12.370 kW

$$\text{flow rate} = \frac{12.37}{4.2 \times 11} = 0.267 \text{ kg/s}$$

From the pipe-sizing Table 9.6, a 28 mm outside diameter tube will carry 0.282 kg/s giving a pressure loss per metre run of 120 Pa. A check may be made on the flow rate through pipes number 9 as follows:

flow rate through pipes number 4 = 0.110 kg/s
flow rate through pipes number 8 = 0.145 kg/s
= 0.255 kg/s

To this flow rate an allowance must be made for the heat emission from pipes number 9.

heat loss from pipes number 9 (5 per cent of 11.781 kW) = 0.589 kW

$$\text{flow rate} = \frac{0.589}{4.2 \times 11} = 0.0127 \text{ kg/s}$$

total flow rate through pipes number 9 = 0.255 + 0.0127
= 0.267 kg/s

This corresponds with the flow rate previously calculated.

Boiler power

The boiler power must be sufficient to supply heat for the radiators, emission from all the pipes, and to heat the water in the storage calorifier.

power required for radiators and pipes = 12.370 kW
power required for hot water supply = 3.000 kW
total = 15.370 kW

A 10 per cent boiler margin may be added for pre-heating and to give a reserve of power for severe weather conditions.

Therefore the boiler power would be:

15.37 kW + 10 per cent = 16.907 kW

A 17 kW boiler would be suitable.

Pump duty

The effective length of the index circuit must first be found by adding an allowance for fittings of 30 per cent on the mains and 10 per cent on the micro-pipe circuit. The index circuit will be the one supplying radiator number 6.

effective length of 10 mm tube

24 m + 10 per cent = 26.4 m
resistance = 26.4 × 1.394 kPa
= 36.96 kPa

effective length of 22 mm and 28 mm tube

7 m + 30 per cent = 9.1 m
resistance = 9.1 × 0.140 kPa
= 1.274 kPa
total resistance = 36.96 + 1.274
= 38.234 kPa

A pump providing a pressure of 40 kPa and giving a mass flow rate of 0.3 kg/s would be satisfactory.

Pipe-sizing for gravity circulation

Gravity-circulation systems are not nowadays used for space-heating systems for anything other than a very small building. The system is, however, still used extensively for the primary circulation between the boiler and calorifier in a hot-water supply installation.

Circulating pressure

The pressure required to ensure circulation in gravity systems is termed 'circulation pressure' and is expressed in pascals per metre of circulating height, i.e. the height measured between the centres of the flow and return connections to the boiler and radiator respectively (see Fig. 9.4).

The circulating pressure per metre of height is found from the following formula:

circulating pressure (Pa per metre height) = $9.81\ (\rho_2 - \rho_1)$

where ρ_2 = density of return water/kg/m^3)
ρ_1 = density of flow water (kg/m^3)

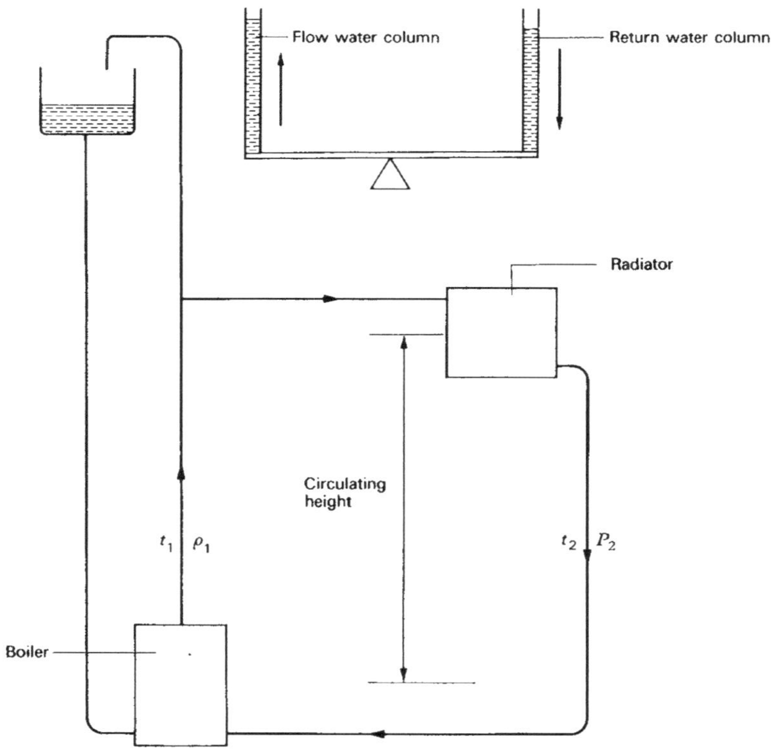

Fig.9.4 Circulation by natural convection

Example 9.4. *Calculate the circulating pressure in pascals when the flow and return water are 82.2 °C and 60 °C respectively and the circulating height is 3m. See Table 9.7 for density of water at various temperatures.*

$$\text{total circulating pressure} = H \times 9.81\ (\rho_2 - \rho_1)$$
$$= 3 \times 9.81\ (983.24 - 970.43)$$
$$= 3 \times 9.81 \times 12.81$$
$$= 376.998\ \text{Pa}$$

This circulating pressure must be sufficient to overcome the pressure due to frictional resistance of the pipe and fittings.

Circulating pressure per metre run

In order to find a suitable diameter of pipe from Table 9.8, it is necessary to find the circulating pressure per metre run of pipe. The run of pipe is the actual length plus the equivalent length due to the resistances of fittings.

Table 9.7 Density of water at various temperatures

Temperature (°C)	Density (kg/m³)
43.3	990.93
48.9	988.56
54.4	986.05
60.0	983:24
65.6	980.29
71.0	977.13
76.7	973.88
82.2	970.43
87.8	966.82
93.3	963.07

Table 9.8 Table for the flow of water at 75 °C in black steel pipes. Heavy-grade pipes to BS 1387

Diameter 20 mm		Diameter 25 mm		Diameter 32 mm			Pressure drop (Pa per m run of pipe)
M	EL	M	EL	M	EL	V	
0.020	0.5	0.037	0.7	0.082	1.0		4.0
0.021	0.5	0.039	0.7	0.087	1.0		4.5
0.022	0.5	0.042	0.7	0.093	1.0		5.0
0.023	0.5	0.044	0.7	0.098	1.0		5.5
0.025	0.5	0.046	0.7	0.103	1.1		6.0
0.026	0.5	0.048	0.7	0.107	1.1		6.5
0.027	0.5	0.050	0.7	0.112	1.1		7.0
0.028	0.5	0.052	0.7	0.116	1.1		7.5
0.029	0.5	0.054	0.7	0.120	1.1		8.0
0.030	0.5	0.056	0.7	0.125	1.1		8.5
0.031	0.5	0.058	0.7	0.129	1.1		9.0
0.032	0.5	0.060	0.7	0.133	1.1		9.5
0.033	0.5	0.062	0.7	0.136	1.1	0.15	10.0
0.037	0.5	0.070	0.8	0.154	1.1		12.0
0.042	0.6	0.077	0.8	0.171	1.2		15.0
0.045	0.6	0.084	0.8	0.186	1.2		17.5
0.049	0.6	0.091	0.8	0.200	1.2		20.0
0.052	0.6	0.097	0.8	0.214	1.2		22.5
0.055	0.6	0.103	0.8	0.226	1.2	.030	25.0

V = velocity of flow m/s
M = mass rate of flow kg/s
EL = equivalent length factor

Example 9.5 (see Fig. 9.5). *Find by use of Table 9.8 the diameter of the primary flow and return pipes for a hot-water cylinder holding a mass of 136 kg of water which is to be raised in temperature from 10 °C to 71 °C in 2 hours. The temperatures of the flow and return pipes are to be 71 °C and 60 °C respectively and the circulating height 1.5 m. An allowance of 20 per cent for heat losses on the pipes and a frictional loss of 10 per cent on the actual length of pipes may be used.*

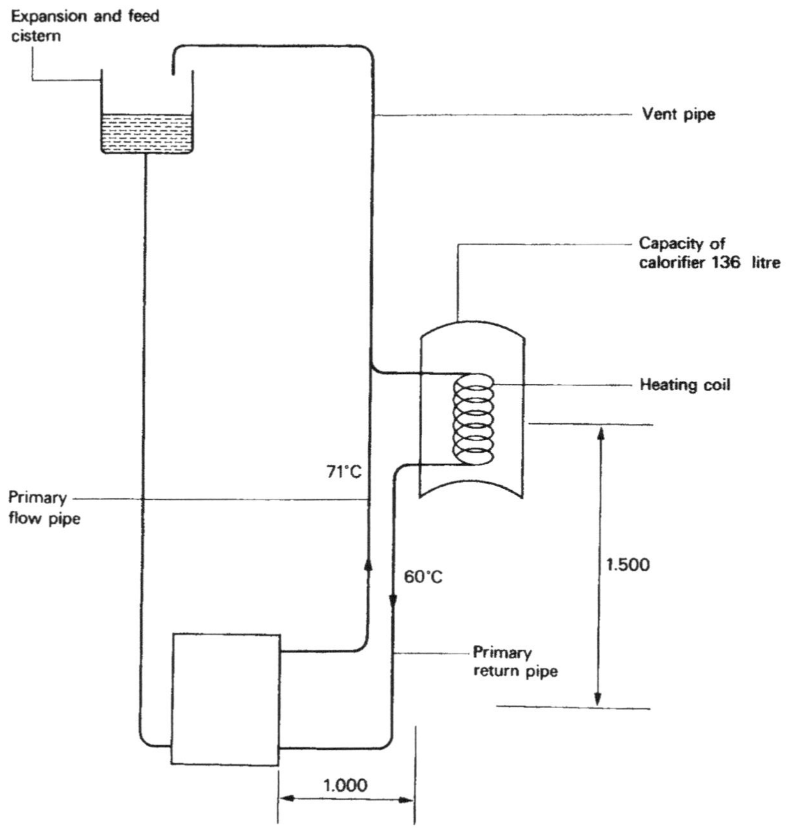

Fig.9.5 Primary circuit

$$\text{power} = \frac{\text{s.h.c.} \times \text{kg} \times (\text{temperature rise } ^\circ\text{C}) \times 100}{3600 \times 2 \times \text{efficiency}}$$

$$= \frac{4.2 \times 136 \times (71-10) \times 100}{3600 \times 2 \times 80}$$

$$= 6.049 \text{ kW}$$

$$\text{flow rate} = \frac{6.049}{4.2 \times 11}$$

$$= 0.131 \text{ kg/s}$$

$$\text{total circulating pressure} = H \times 9.81\ (\rho_2 - \rho_1)$$

$$= 1.5 \times 9.81 \times (983.24 - 977.13)$$

$$= 1.5 \times 9.81 \times 6.11$$

$$= 90 \text{ Pa}$$

$$\text{effective length of pipe} = 5 + 10 \text{ per cent}$$

$$= 5.5 \text{ m}$$

$$\text{circulating pressure available per metre run} = \frac{\text{total circulating pressure}}{\text{effective length}}$$

$$= \frac{90}{5.5}$$

$$= 16.36 \text{ Pa}$$

From the pipe-sizing Table 9.8, a 32 mm diameter steel pipe will carry 0.133 kg/s giving a pressure loss of 9.5 Pa per metre run. Since the flow rate for this pipe is greater than required and the frictional loss is less, the pipe will be satisfactory.

Sizing of secondary circuits

The secondary flow pipe will have to deliver a certain rate of flow to the various sanitary fittings, while the secondary return will have to carry sufficient water to compensate for the heat losses from the pipes and any heat emitted from towel rails, etc. The secondary return may therefore be determined as previously described for the pipe-sizing for heating.

The sizing of the secondary flow pipe can be carried out from charts or tables or by the Box formula described in Chapter 3, using the discharge rates from the sanitary fittings as listed in Table 9,9, and allowing a diversity factor.

Table 9.9 Recommended minimum rate of flow at various appliances

Type of appliance	Rate of flow (litre/s)
WC flushing cistern	0.12
Wash basin	0.15
Wash basin with spray taps	0.04
Bath (private)	0.30
Bath (public)	0.60
Shower (with nozzle)	0.12
Sink with 13 mm taps	0.20
Sink with 19 mm taps	0.30
Sink with 25 mm taps	0.60

Index circuit (see Fig. 9.6)

In a gravity-heating system the index circuit may be defined as the circuit in which the ratio of the circuit length to the circuit circulating height is the minimum.

If the temperature drop across each circuit in Fig. 9.6 is the same, and since circuit 1 has twice the circulating height of circuit 2, it will have double the circulating pressure of circuit 2. Since circuit 2 has a greater length than circuit 1, it follows that circuit 2 is the index circuit.

In order to find the diameters of pipes for the system, the circulating pressure per metre of travel will be based on circuit 2. The method of pipe-sizing will then be carried out as described for the primary flow and return pipes in a hot-water supply system.

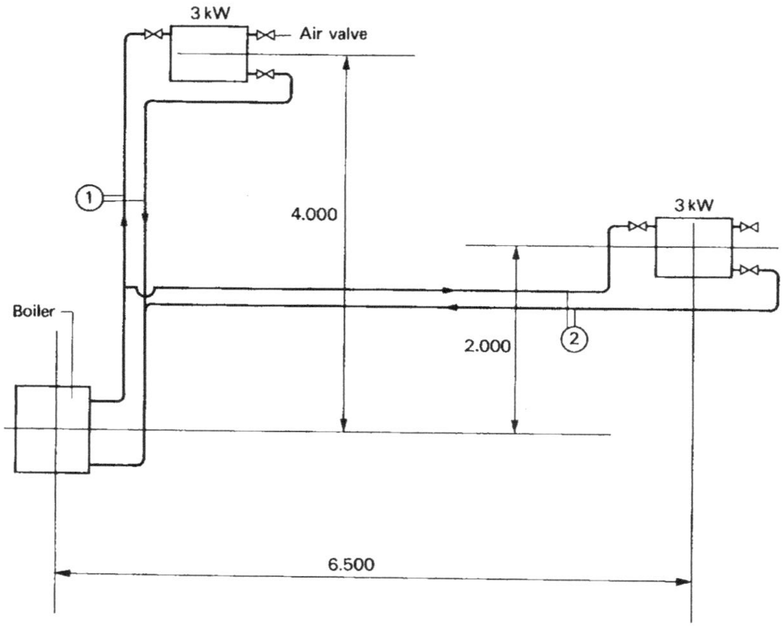

Fig.9.6 Index circuit

Questions

1. Figure 9.7 shows a single-pipe circuit, assuming each radiator has a heat emission of 5 kW and the temperature drop across the system is 10 °C. Find by use of Table 9.10 the approximate diameter of a copper main pipe. A velocity of flow of water is to be 0.5 m/s. Determine the pump duty for the circuit allowing an emission from the pipes 20 per cent of the total from the radiators.

Answers: 51 mm diameter; pump duty – pressure 3 kPa; discharge 0.86 kg/s (0.9 kg/s approx.)

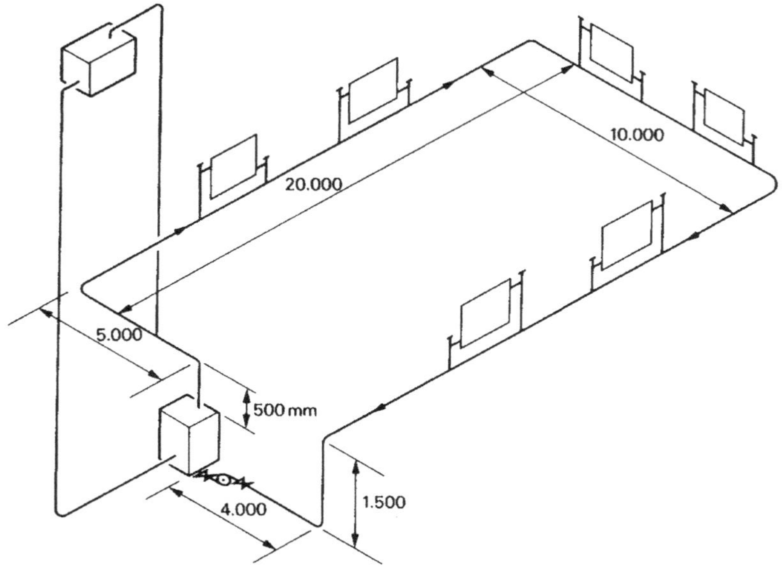

Fig 9.

Table 9.10 Table for the flow of water at 75 °C in copper tubes BS 659 Light Gauge

Diameter 51 mm		Diameter 63 mm		Diameter 76 mm			Pressure drop (Pa per m run of pipe)
M	EL	M	EL	M	EL	V	
0.428	2.2	0.774	2.9	1.26	3.8	0.30	10.0
0.486	2.2	0.878	3.0	1.42	3.8		12.5
0.538	2.3	0.972	3.1	1.58	3.9		15.0
0.587	2.3	1.06	3.1	1.72	4.0		17.5
0.633	2.4	1.14	3.2	1.85	4.1		20.0
0.676	2.4	1.22	3.2	1.98	4.1		22.5
0.714	2.4	1.29	3.3	2.10	4.2		25.0
0.757	2.5	1.37	3.3	2.21	4.2		27.5
0.795	2.5	1.43	3.3	2.32	4.3		30.0
0.838	2.5	1.50	3.4	2.42	4.3		32.5
0.867	2.5	1.56	3.4	2.53	4.3		35.0
0.907	2.5	1.62	3.4	2.63	4.4	0.50	37.5

V = velocity of flow m/s
M = mass rate of flow kg/s
EL = equivalent length factor

2. Figure 9.8 shows a two-pipe circuit assuming each radiator has a heat emission of 8 kW and the temperature drop across the system is 10 °C. Find by use of Table 9.11 the approximate diameters of the steel pipes. Determine the pump duty for the circuit allowing an emission from each section of the pipework 5 per cent of the total from the radiators.

Answers: pipe number 1, 20 mm diameter; pipe number 2, 25 mm diameter; pump duty – pressure 5 kPa; discharge 0.3 kg/s.

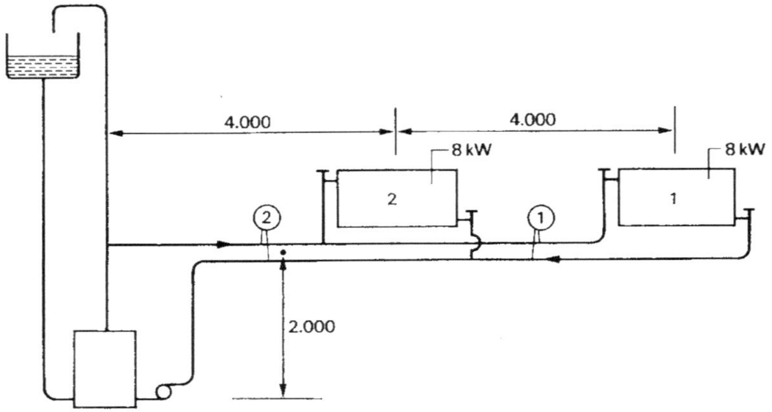

Fig 9.8

Table 9.11 Table for the flow of water at 75 °C in black steel pipes. Heavy-grade pipes to BS 1387

Diameter 15 mm		Diameter 20 mm		Diameter 25 mm			Pressure drop (Pa per m run of pipe)
M	EL	M	EL	M	EL	V	
0.011	0.3	0.026	0.5	0.048	0.7	0.15	6.5
0.011	0.3	0.027	0.5	0.050	0.7	0.15	7.0
0.012	0.3	0.028	0.5	0.052	0.7	0.15	7.5
0.012	0.3	0.029	0.5	0.054	0.7	0.15	8.0
0.013	0.3	0.030	0.5	0.056	0.7	0.15	8.5
0.013	0.3	0.031	0.5	0.058	0.7	0.15	9.0
0.014	0.3	0.032	0.5	0.060	0.7	0.15	9.5
0.014	0.3	0.033	0.5	0.062	0.7	0.15	10.0
0.016	0.3	0.037	0.5	0.070	0.8	0.15	12.5
0.018	0.4	0.042	0.6	0.077	0.8	0.15	15.0
0.019	0.4	0.045	0.6	0.084	0.8	0.15	17.5
0.021	0.4	0.049	0.6	0.091	0.8	0.15	20.0

V = velocity of flow m/s
M = mass rate of flow kg/s
EL = equivalent length factor

3. Define the following terms: (*a*) index circuit; (*b*) index radiator; (*c*) circulating pressure; (*d*) circulating height.

4. Figure 9.9 shows a part of a micro-pipe system which represents the index circuit. If an 'open' type of system is to be used with flow and return water temperatures of 80 °C and 69 °C respectively, find the diameters of the micro-pipes and the total pressure drop in pascals through them.

Answers: 10 mm O.D. copper tube; pressure drop 24 805 Pa or 24.8 kPa.

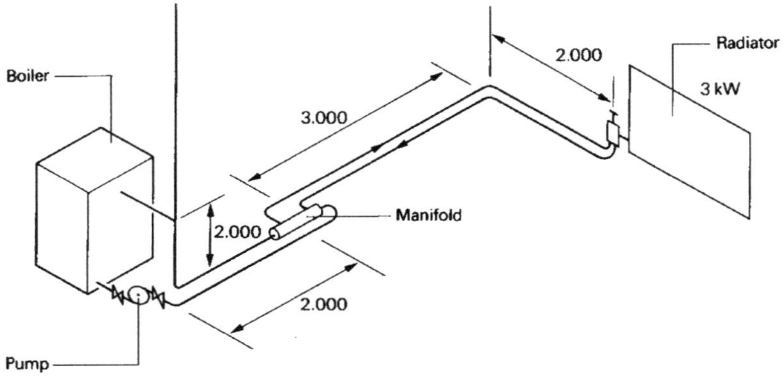

Fig 9.9

5. Calculate the total circulating pressure for a gravity heating system when the circulating height is 4 m and the flow and return waters are 82.2 °C and 60 °C respectively.

Answer: 502.3 Pa

6. Find by the use of Table 9.8 suitable diameter flow and return pipes for a 136 litre hot-water cylinder. Use the following factors:

(*a*) heat recovery period = 3 hours
(*b*) circulating height = 1.2 m
(*c*) actual length of pipes = 4.4 m
(*d*) temperature flow of water = 71 °C
(*e*) temperature of return water = 60 °C
(*f*) temperature rise of water in the cylinder = 65 °C
(*g*) heat loss in pipes = 20 per cent of power required to heat water
(*h*) frictional loss on pipes = 10 per cent of actual length

Answer: 32 mm diameter steel pipe

Chapter 10

Ventilating ducts and fans

Ventilating ducts

The sizing of ducts to convey air from fans in a plant room to the various spaces inside a building relies upon empirical data for the evaluation of frictional losses. It cannot therefore be assumed that the calculated volumes of air will correspond with those actually delivered to the various discharge points, and for this reason dampers are incorporated inside the duct system. These dampers perform the same function as regulating valves in hot-water heating systems.

Flow of air in ducts

The Bernoulli theorem may be applied to air flow in ducts provided that an allowance is made for friction and separation. The theorem states that the total energy of each particle of a fluid is the same, provided that no energy enters or leaves the system at any point. If there is a loss of one type of energy, there must be a corresponding gain in another, or vice versa.

Figure 10.1 shows the total energies of air flowing inside a section of ductwork. The total energy possessed by the air is the sum of the potential, pressure and kinetic energies and this can be expressed as follows:

$$\text{total energy } H = z + \frac{P}{\rho g} + \frac{V^2}{2g} = \text{constant}$$

where z = potential energy due to the position of the particles or height above the datum (m)

$\frac{P}{\rho g}$ = pressure energy due to the depth of the particles below the fluid (m)

$\frac{V^2}{2g}$ = kinetic energy due to the movement of particles of the fluid (m)

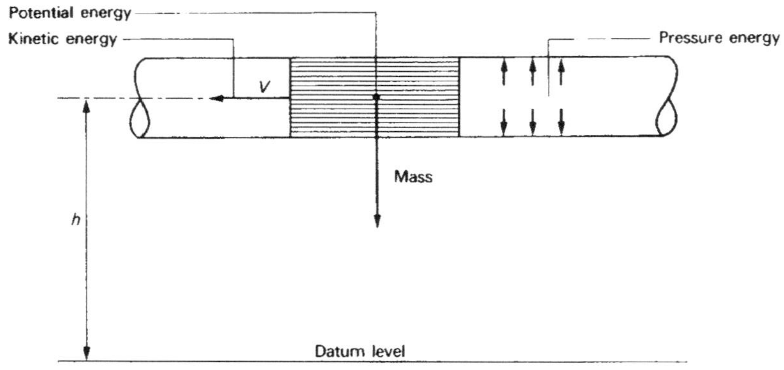

Fig. 10.1 Total energy of a moving fluid

Because the mass of air involved and its density are small, the value of the potential energy will also be small and for practical purposes is neglected. This reduces the equation to:

$$H = \frac{P}{\rho g} + \frac{V^2}{2g} = \text{constant}$$

Pressure distribution

In a closed system the total energy at two points must be equal, provided that there are no losses or gains.

$$\frac{P_1}{\rho g} + \frac{V_1{}^2}{2g} = \frac{P_2}{\rho g} + \frac{V_2{}^2}{2g}$$

If ideal conditions existed, i.e. no pressure loss due to friction, the distribution of pressure would be as shown in Fig. 10.2.

The loss of pressure energy due to friction between two points in a duct may be described in diagrammatic form (see Fig. 10.3).

The pressure energy $P/\rho g$ is referred to as the static head, or pressure of the system, and may be defined as the pressure acting equally in all directions. It is the pressure that tends either to burst the duct outwards when it is positive, or to collapse the duct inwards if it is negative.

The velocity energy $V^2/2g$ is referred to as the velocity head, or pressure of the system, and may be defined as the energy required either to accelerate the air from rest to a certain velocity, or to bring the air from a certain velocity to rest.

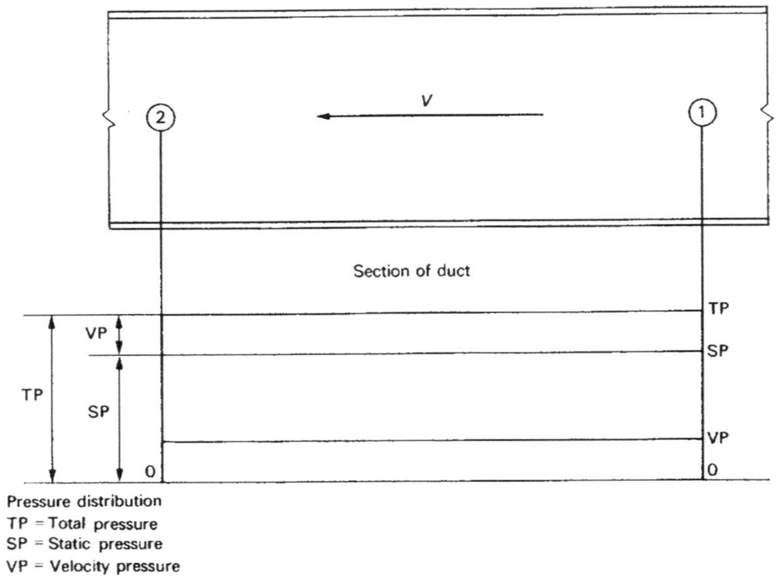

Fig. 10.2 Total energy at two points in a ventilating duct for ideal conditions

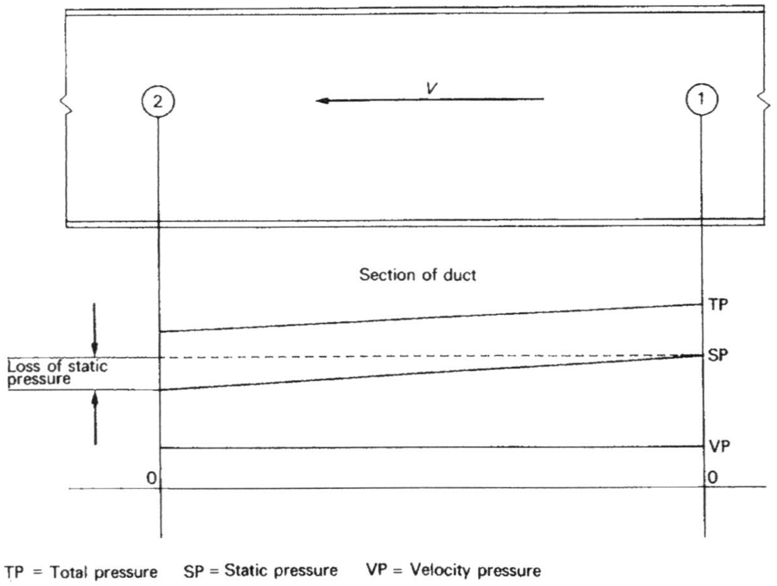

Fig. 10.3 Total energy at two points in a ventilating actual duct for conditions (i.e. pressure loss due to friction)

Measurement of pressure (see Fig. 10.4)

In order to measure the pressures inside a duct, a Pitot tube is used. One manometer will show the total pressure while the other will show the static pressure.

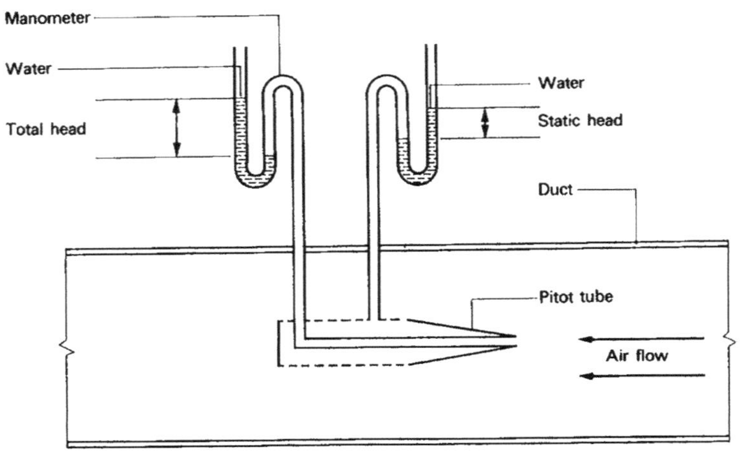

Fig. 10.4 Pitot tube

The velocity pressure may be found by deducting the static pressure from the total pressure, i.e.

$$\text{velocity pressure or head} = \text{total pressure or head} - \text{static pressure or head}$$

Because the manometer-tube readings are very small, in order to obtain a higher degree of accuracy, an inclined manometer may be used with the Pitot tube.

The velocity pressure is always positive and since the total pressure is the sum of the static and velocity pressure, the total pressure may therefore be either positive or negative.

Velocity and velocity head (VH)

$$\text{VH} = \frac{V^2}{2g}$$

since the measurement is expressed in water gauge, and air is being considered:

$$\text{VH} = \frac{V^2}{2g} \times \frac{\rho_a}{\rho_w}$$

where VH = velocity head (m)
V = velocity of air flow (m/s)
ρ_a = density of air (1.2 kg/m^3)
ρ_w = density of water (998 kg/m^3)

$$\therefore \text{VH} = \frac{V^2}{2 \times 9.81} \times \frac{1.2}{998}$$

$$\text{and} = V^2 \times 0.000\,061\,2 \text{ (m) of water}$$

In terms of mm of water,

$$\text{VH} = V^2 \times 0.0612$$

and

$$V = \frac{\text{VH}}{0.0612}$$

Velocity pressure (VP)

Manufacturers of fans often state the pressure developed by a fan in pascals instead of in millimetres water gauge. The pressure in pascals may be found as follows:

$$\text{pressure (Pa)} = \text{head of water (m)} \times 1000 \times 9.81$$
$$\therefore \text{VP} = V^2 \times 0.000\,061\,2 \times 1000 \times 9.81$$
$$= V^2 \times 0.6 \text{ Pa (approx.)}$$

This is the form given in the Chartered Institution of Building Services guide book and a table listing velocity pressure in pascals against velocity in m/s is produced.

Example 10.1. *Calculate the velocity pressure in pascals in a ventilating duct when the velocity of air is found to be 5 m/s.*

$$\text{VP} = 5^2 \times 0.6$$
$$= 15 \text{ Pa}$$

Example 10.2. *If the velocity pressure of air flowing in a duct is found to be 10 pascals, calculate the velocity of the air.*

$$\text{VP} = V^2 \times 0.6$$

$$V = \sqrt{\frac{\text{VP}}{0.6}}$$

$$V = 4 \text{ m/s (approx.)}$$

Example 10.3. *Calculate the velocity head in mm water gauge when the velocity of flow of air in a duct is 4 m/s.*

$$\text{VH} = 4^2 \times 0.0612$$
$$= 0.979 \text{ mm (approx.)}$$

Example 10.4. *Calculate the theoretical velocity of the flow of air in a duct when the velocity head is found to be 4 mm water gauge.*

$$V = \sqrt{\frac{4}{0.0612}}$$
$$= 8 \text{ m/s (approx.)}$$

Note: In practice, the velocity of air flow in a duct is often found by means of a velometer or anemometer.

Volume of air flow

The following formula may be used:

$$Q = VA$$

where Q = volume of air flow (m^3/s)
V = velocity of air flow (m/s)
A = cross-sectional area of duct (m^2)

Example 10.5. *The flow rate in a circular duct is to be determined by means of a Pitot tube. Determine the volume of flow in m^3/s from the following data:*

diameter of duct = *600 mm*
total head (TH) = *30 mm water gauge*
static head (SH) = *18 mm water gauge*
standard air density = *1.20 kg/m^3 at 20°C*

$$\text{VH} = \text{TH} - \text{SH}$$
$$= 30 - 18$$
$$= 12 \text{ mm water gauge}$$

$$V = \sqrt{\frac{12}{0.0612}}$$

$$V = 14 \text{ m/s}$$
$$Q = VA$$
$$Q = V \times 0.7854\,D^2$$
$$Q = 14 \times 0.7854 \times 0.600 \times 0.600$$
$$Q = 3.96 \text{ m}^3\text{/s (approx.)}$$

Example 10.6. *A room measuring 20 m × 10 m × 3 m requires ventilating by means of a fan and ductwork to provide six air changes in the room. If the average velocity of air flow in the duct is to be 2 m/s, calculate the diameter of the main circular duct required for the room.*

$$Q = \frac{20 \times 10 \times 3 \times 6}{3600}$$

$$Q = 1 \text{ m}^3\text{/s}$$

$$Q = V \times 0.7854\,D^2$$

$$D = \sqrt{\frac{Q}{V \times 0.7854}}$$

$$D = \sqrt{\frac{1}{2 \times 0.7854}}$$

$$D = 0.798 \text{ m or } 798 \text{ mm}$$

The D'Arcy formula

It was shown in Chapter 3 that the loss of head due to friction for water flowing through a pipe was given in the following expression:

Example 10.10. *Determine the total loss of head in mm water gauge in a 300 mm diameter duct 15 m long, having three 90° bends when the velocity of air flow is 6 m/s.*

$$\text{velocity-head loss due to bends} = 3 \times 0.1 \left[\frac{6^2}{2 \times 9.81}\right] \times 0.0012$$
$$= 0.000\,66 \text{ m}$$
$$= 0.66 \text{ mm}$$

Static-head loss due to straight duct, using the duct-sizing chart (Fig. 10.5) for 300 mm diameter and 6 m/s.

$$\text{for } 1 \text{ m} = 0.19 \text{ mm}$$
$$\text{for } 15 \text{ m} = 0.19 \times 15$$
$$= 2.85 \text{ mm}$$
$$\text{total head loss} = 0.66 + 2.85$$
$$= 3.51 \text{ mm}$$

Sizing of system of ductwork

Several methods may be used to determine the sizes of ventilating ducts to deliver or extract air from various spaces inside a building. In theory, duct sizing is very complex, but the following approximate methods are used in practice:

Equal-velocity method: The air velocity in the ductwork is decided by the designer and duct cross-sectional areas obtained from the volumes of flows required using the continuity equation. This method is generally used for simple systems without branches and for pneumatic conveying.

Velocity-reduction method: The air velocity in the first section of the main duct is chosen and the velocity reduced at each branch. Duct sizes are then found using the continuity equation.

Static-regain method: The size of a duct between two adjacent branches is chosen so that the pressure drop over the section is equal to the static regain at the previous branch. Therefore the static pressure at every branch is the same.

Constant pressure drop per unit length: The air velocity in the first section of the main duct is chosen and the size of the duct in this section found. Using the duct-sizing chart, the pressure drop per unit length of duct is obtained for this section and this value is then used for all the remaining sections of ductwork. This is a very convenient method of duct sizing and will be the one used in the following example.

For low-velocity systems a total pressure drop of 1 Pa may be used.

Example 10.11. *Figure 10.7 shows an extract system of ductwork. Determine by use of the duct-sizing chart (Fig. 10.5) the diameters of the ducts A, B and C assuming that the average velocity of air flow through duct A is to be 5 m/s.*

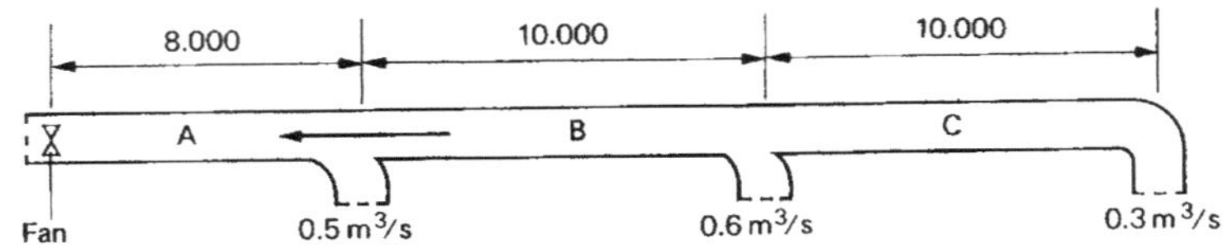

Fig. 10.7 System of ductwork

Duct A

$$\text{volume of flow} = 0.5 + 0.6 + 0.3$$
$$= 1.4 \text{ m}^3/\text{s}$$

From the duct-sizing chart a flow rate of 1.4 m^3/s and an air velocity of 5 m/s gives a head loss per metre run of 0.059 mm water gauge and a 600 mm diameter duct.

Duct B

Using the same head loss per metre run as for duct A,

$$\text{flow rate} = 0.6 + 0.3$$
$$= 0.9 \text{ m}^3/\text{s}$$

From the duct-sizing chart a 500 mm diameter duct giving an air velocity of approximately 4.4 m/s is found to be suitable.

Duct C

Using the same head loss per metre run as for duct A,

$$\text{flow rate} = 0.3 \text{ m}^3/\text{s}$$

From the duct-sizing chart a 350 mm diameter duct giving an air velocity of approximately 3.4 m/s is found to be suitable.

Total head loss

The values of resistances through the various fittings and the lengths of ductwork can now be calculated basing the velocity head on the velocity through each section of ductwork. The total loss of pressure through the duct will be required when ordering the fan.

Duct A

$$\text{loss of head through straight duct} = 8 \times 0.059$$
$$= 0.422 \text{ mm}$$
$$= 0.000\,422 \text{ m}$$

To this must be added the loss of head through the outlet, which from Table 10.1 has a k factor of 1.0.

$$\text{loss through outlet} = 1 \left[\frac{5^2}{2 \times 9.81}\right] \times 0.0012$$
$$= 0.0015 \text{ m}$$
$$\text{total loss} = 0.000\,422 + 0.0015$$
$$= 0.001\,922 \text{ m}$$
$$= 0.002 \text{ m (approx.)}$$

Duct B

$$\text{loss of head through straight duct} = 10 \times 0.059 = 0.59 \text{ mm} = 0.000\,59 \text{ m}$$

To this must be added resistances of two branches and two inlets.

$$\text{branches (average } k\text{, 0.9)} = 2 \times 0.9 \left[\frac{4.4^2}{2 \times 9.81}\right] \times 0.0012$$

$$= 2 \times 0.9 \times 0.987 \times 0.0012$$

$$= 0.002\,13 \text{ m}$$

$$\text{inlets (wire mesh } k\text{, 0.4)} = 2 \times 0.4 \left[\frac{4.4^2}{2 \times 9.81}\right] \times 0.0012$$

$$= 2 \times 0.4 \times 0.987 \times 0.0012$$

$$= 0.000\,947 \text{ m}$$

$$\text{total loss} = 0.000\,59 + 0.002\,13 + 0.000\,947$$

$$= 0.003\,667 \text{ m}$$

$$= 0.004 \text{ m (approx.)}$$

Duct C

$$\text{loss of head through straight duct} = 10 \times 0.059 = 0.59 \text{ mm} = 0.000\,59 \text{ m}$$

To this must be added one bend and one inlet.

$$\text{bend } (k, 0.1) = 0.1 \left[\frac{3.4^2}{2 \times 9.81}\right] \times 0.0012$$

$$= 0.1 \times 0.589 \times 0.0012$$

$$= 0.000\,070\,6 \text{ m}$$

$$\text{inlet } (k, 0.4) = 0.4 \left[\frac{3.4^2}{2 \times 9.81}\right] \times 0.0012$$

$$= 0.4 \times 0.589 \times 0.0012$$

$$= 0.000\,287 \text{ m}$$

$$\text{total loss} = 0.000\,59 + 0.000\,07 + 0.000\,287$$

$$= 0.000\,947 \text{ m}$$

$$\text{loss through duct system} = 0.002 + 0.004 + 0.0095$$

$$= 0.0155 \text{ m}$$

Converting this loss of head to pressure (Pa)

$$\text{pressure} = 0.0155 \times 1000 \times 9.81$$

$$= 152 \text{ Pa}$$

Fan duty

A fan that will discharge 1.4 m^3/s and develop a total negative pressure of 152 Pa would be suitable. Dampers placed in the inlet branches will enable the system to be balanced.

Fans

There are three main types of fans used for ventilation systems:

Propeller fan: This type is usually used at free opening in walls or windows and for other types of low pressure applications. It is not usually employed for ducted ventilation systems.

Axial-flow fan: This type of fan is designed for mounting inside a duct system and is suitable for moving air in complete systems of ductwork.

Centrifugal fan: The inlet of the fan is at 90° to the outlet and like the axial-flow fan is suitable for moving air in complete systems of ductwork.

Fan laws

The performance of a fan incorporated in a system of ventilation is governed by the following laws, providing the air density remains constant:

1. The discharge varies directly with the angular velocity of the impeller.
2. The pressure developed varies directly as the square of the angular velocity of the impeller.
3. The power absorbed varies as the cube of the angular velocity of the impeller.

The laws may be expressed as follows:

1. $$\frac{Q_2}{Q_1} = \frac{N_2}{N_1}$$
2. $$\frac{P_2}{P_1} = \frac{(N_2)^2}{(N_1)^2}$$
3. $$\frac{\text{power}_2}{\text{power}_1} = \frac{(N_2)^3}{(N_1)^3}$$

where Q = volume of flow in m^3/s

N = revolutions of impeller per minute

P = pressure in pascals

power = power in watts or kilowatts

Example 10.12. *A fan absorbs 2.3 kW of power and discharges 2.5 m^3/s when the impeller angular velocity is 1000 revolutions per minute. If the impeller angular velocity is increased to 1200 revolutions per minute, calculate the discharge in m^3/s and the power absorbed for this new condition.*

$$\frac{Q_2}{Q_1} = \frac{N_2}{N_1}$$

$$\therefore Q_2 = \frac{Q_1 \times N_2}{N_1}$$

$$Q_2 = \frac{2.5 \times 1200}{1000}$$

$$= 3 \text{ m}^3/\text{s}$$

$$\frac{\text{power}_2}{\text{power}_1} = \frac{(N_2)^3}{(N_1)^3}$$

$$\therefore \text{power}_2 = \frac{W_1 \times (N_2)^3}{(N_1)^3}$$

$$= \frac{2.3 \times 1200^3}{1000^3}$$

$$= 3.9744 \text{ kW}$$

It will be observed from example 10.12 that for a 20 per cent increase in angular velocity there is a 72.8 per cent increase in power absorbed. The common method used to lower the air flow through a main duct is by closing a damper, and this can lead to a significant loss of power. An alternative method of lowering the air flow is by lowering the angular velocity of the impeller. This can be achieved by varying the frequency of the current input to the motor driving the fan.

Example 10.13. *A fan develops a static pressure of 200 Pa when the angular velocity of the impeller is 900 revolutions per minute. If the angular velocity of the impeller is increased to 1000 revolutions per minute, calculate the static pressure developed by the fan for the new condition.*

$$\frac{P_2}{P_1} = \frac{(N_2)^2}{(N_1)^2}$$

$$\therefore P_2 = \frac{P_1 \times (N_2)^2}{(N_1)^2}$$

$$= \frac{200 \times 1000^2}{900^2}$$

$$= 246.9$$

$$= 247 \text{ Pa (approx.)}$$

Efficiency of fans

The efficiency of a fan can be found by use of the following expressions:

$$\text{efficiency (per cent)} = \frac{\text{fan total pressure} \times \text{volume of flow}}{\text{power absorbed (W)}} \times \frac{100}{1}$$

Example 10.14. *A fan absorbs 1.5 kW when discharging 2 m³/s of air and operating at a total pressure of 500 Pa. Calculate the percentage efficiency of the fan.*

$$\text{efficiency} = \frac{500 \times 2}{1500} \times \frac{100}{1}$$

$$= 66.667 \text{ per cent}$$

Change of air density

If the air density is changed, the following laws apply:

1. The volume of flow remains constant.
2. The pressure developed varies directly with the change in density.
3. The power absorbed varies directly with the change in density.

The laws may be expressed as follows:

1. $Q_1 = Q_2$
2. $\dfrac{P_2}{P_1} = \dfrac{\rho_2}{\rho_1}$
3. $\dfrac{\text{power}_2}{\text{power}_1} = \dfrac{\rho_2}{\rho_1}$

where Q = volume of flow in m³/s
P = pressure in pascals, which may be static, velocity or total
ρ = density of air in kg/m³
power = power in watts or kilowatts

Example 10.15. *A fan develops a total pressure of 400 Pa when discharging air at a temperature of 20 °C. If the temperature of the air is lowered to 16 °C, calculate the pressure to be developed by the fan in order to discharge the same volume of air.*

density of air at 20 °C = 1.2 kg/m³
density of air at 16 °C = 1.22 kg/m³

$$\frac{P_2}{P_1} = \frac{\rho_2}{\rho_1}$$

$$\therefore P_2 = \frac{P_1 \times \rho_2}{\rho_1}$$

$$= \frac{400 \times 1.22}{1.2}$$

$$= 406.67 \text{ Pa}$$

Selection of a fan

In order to select a fan for a given duty, reference should be made to fan-performance graphs supplied by the manufacturer.

Figure 10.8 shows a typical performance graph for two fans and it will be seen that the following factors may be obtained from them:

1. pressure in pascals;
2. volume flow rate in m³/s;
3. input in watts.

Before using the graphs it is essential to draw a system characteristic curve and the intersection of this curve with the fan-performance curve will give the actual amount of discharge and the pressure developed by the fans.

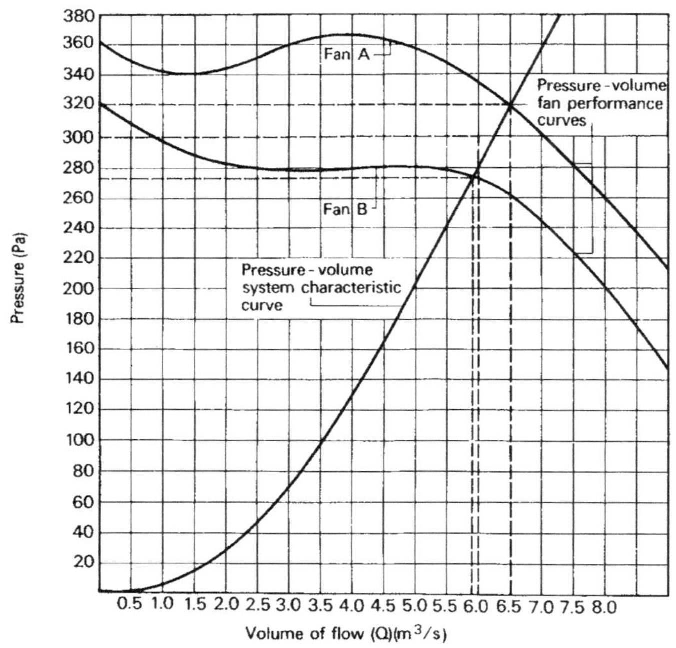

Note: Manufacturers of fans will supply graphs giving pressure, volume of air flow, power and efficiency

Fig.10.8 Performance graph for fans

System characteristic curve

For a certain rate of flow of air through a system of ductwork a certain static pressure must be developed.

For any system of ductwork therefore, a characteristic curve may be drawn plotting the rate of flow against the static pressure to be developed in order to overcome the resistances.

Example 10.16. *A ventilating system requires a fan to discharge 6 m^3/s against a calculated resistance of 300 Pa pressure. Select from the graph in Fig. 10.8 either fan A or fan B.*

For a given system of ductwork the pressure loss may be found from the following expression:

pressure loss = pressure-loss coefficient × volume of flow²

$$P = k\,Q^2$$

$$\therefore\ k = \frac{P}{Q^2} = \frac{300}{6^2} = 8.333$$

In order to draw the system characteristic curve it is necessary to find the values of pressure corresponding to nominated values of flow. Using the flow rates between 0.5 and 6 m^3/s :

1. $Q = 8.33 \times 0.5^2 =$ 2.08 Pa
2. $Q = 8.33 \times 1.0^2 =$ 8.33 Pa
3. $Q = 8.33 \times 1.5^2 =$ 18.74 Pa
4. $Q = 8.33 \times 2.0^2 =$ 33.32 Pa
5. $Q = 8.33 \times 2.5^2 =$ 52.06 Pa
6. $Q = 8.33 \times 3.0^2 =$ 74.97 Pa
7. $Q = 8.33 \times 3.5^2 =$ 102.04 Pa
8. $Q = 8.33 \times 4.0^2 =$ 133.28 Pa
9. $Q = 8.33 \times 4.5^2 =$ 168.68 Pa
10. $Q = 8.33 \times 5.0^2 =$ 208.25 Pa
11. $Q = 8.33 \times 5.5^2 =$ 251.98 Pa
12. $Q = 8.33 \times 6.0^2 =$ 299.88 Pa

By reference to Fig. 10.8, fan A will discharge 6.5 m^3/s and develop a pressure of 320 Pa for the conditions given. Fan B would discharge 5.9 m^3/s and develop a pressure of 272 Pa for the same conditions.

The designer has a choice between fan A and fan B, and fan A would probably be chosen. If it is required to save electrical power, however, fan B might be selected.

Limiting velocities in ducts

In order to reduce noise and power consumption, the velocity of air flowing through a duct must be kept within reasonable limits. Table 10.2 gives values of limiting velocities in ducts.

Table 10.2 Limiting velocities in ducts

Application	Velocity (m/s)	
	Commercial buildings	Industrial buildings
Outside air intake	5.0	7.5
Discharge to atmosphere	5.0	7.5
Main supply or extract duct	7.5	13.0
Terminal branch duct	3.5	5.0

Questions

1. Make a statement of Bernoulli's theorem as it applies to air flow in ductwork.
2. Define velocity, static and total pressure and describe how the Pitot tube may be used to measure the velocity, static and total pressures of air flowing through a duct.

3. Calculate the velocity pressure in pascals in a ventilating duct when the velocity of air is found to be 7 m/s.

Answer: 29.4 Pa

4. Calculate the velocity of air flowing in a duct when the velocity pressure is 25 Pa.

Answer: 6.455 m/s

5. Calculate the flow rate in m^3/s through a 350 mm diameter ventilating duct when the total and static heads are 30 mm water gauge and 25 mm water gauge respectively. Assume standard air density of 1.20 kg/m^3.

Answer: 0.87 m^3/s (approx.)

6. A room measuring 15 m × 8 m × 3 m requires ventilating by means of a fan and ductwork to provide three air changes per hour. If the average velocity of air flow in the duct is to be 4 m/s, calculate the diameter of the main circular duct required for the room.

Answer: 309 mm

7. Calculate the static head lost due to friction in a 300 mm diameter ventilating duct 15 m long when the average velocity of air flow through the duct is 5 m/s. Use the following values:

(*a*) density of water = 998 kg/m^3
(*b*) density of air = 1.2 kg/m^3
(*c*) coefficient of friction = 0.005

Answer: 1.53 mm water gauge

8. Calculate the static head lost due to friction in a rectangular ventilating duct having sides 350 mm by 300 mm. The length of the duct is 12 m and the average velocity of air flowing through it 3 m/s. Use the values given in Question 7.

Answer: 0.8 mm water gauge

9. Find by use of the duct-sizing chart (Fig. 10.5) the diameter of a circular duct 50 m long that will give a flow rate of 9 m^3/s when the velocity of flow is 5 m/s. Find the total static head lost due to friction in mm water gauge.

Answers: diameter 1.4 m; total static head lost 0.9 mm water gauge

10. Determine the total loss of head in mm water gauge in a 500 mm diameter duct 30 m long, having four 90° bends, when the velocity of air flow is 5 m/s.

Answer: 2.8 mm water gauge (approx.)

11. Figure 10.9 shows an inlet system of ductwork. Determine by use of the duct-sizing chart (Fig. 10.5) the diameters of the ducts A and B. It may be assumed that the average velocity of air flow through duct A is to be 6 m/s.

Answers: duct A 500 mm diameter; duct B 450 mm diameter

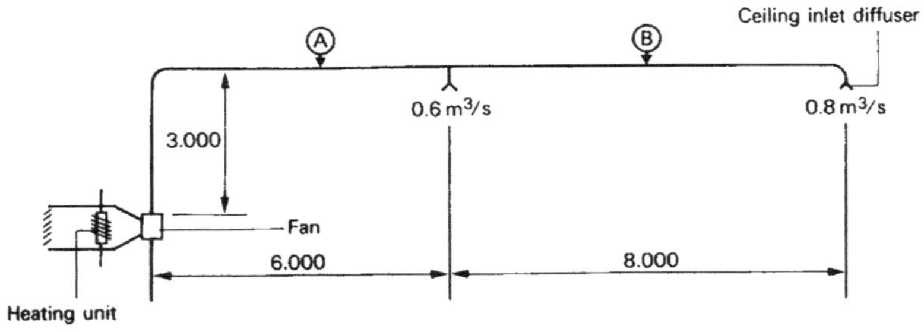

Fig. 10.9

12. State the fan laws for: (*a*) constant air density; (*b*) variable air density.

13. What information is required to be given to a manufacturer when ordering a fan for a certain system of ventilation?

14. A fan absorbs 1.25 kW of power and discharges 2.3 m^3/s when the impeller angular velocity is 1300 revolutions per minute. If the impeller is reduced to 1000 revolutions per minute, calculate the discharge in m^3/s and the power absorbed for this new condition.

Answers: 1.769 m^3/s; 0.569 kW

15. A fan absorbs 1.5 kW when discharging 3 m^3/s of air and operating at a total pressure of 400 Pa. Calculate the percentage efficiency of the fan.

Answer: 80 per cent

16. A fan develops a total pressure of 500 Pa when discharging air at a density of 1.2 kg/m^3. If the temperature of the air is raised so that the air density becomes 0.95 kg/m^3, calculate the pressure to be developed by the fan in order to discharge the same volume of air.

Answer: 395.833 Pa

Chapter 11

Heat gains, solar heating

Solar heat gains (principles)

In order to understand calculations involved in determining heat gains in a building due to solar radiation, a knowledge of the relative movements of the earth and sun are essential.

The earth rotates about its axis once every 24 hours and revolves about the sun once every year. The axis of the earth is inclined at an angle of 23.5° to the ecliptic. At the winter solstice, i.e. 21 December, the North Pole is inclined away from the sun and at the summer solstice, i.e. 21 June, the North Pole is inclined towards the sun. These changes are due to movement of the earth in its orbit around the sun and the two positions are shown in Fig. 11.1.

By inspection of Fig. 11.1 it will be noticed that the sun will be seen directly overhead at different earth latitudes. On 21 December, the sun will be seen directly overhead on the Tropic of Capricorn and on 21 June, it will be directly overhead on the Tropic of Cancer.

Countries to the south of the Tropic of Capricorn and to the north of the Tropic of Cancer will, however, never have the sun appearing directly overhead. At the spring and autumn equinoxes, i.e. 21 March and 21 September, the earth's poles are equidistant from the sun, so that everywhere on the earth's surface has 12 hours of daylight and 12 hours of darkness. The sun also appears directly overhead at the equator on these days (see Fig. 11.2).

Monthly declination

Since four fixed positions of the earth relative to the sun are known, it is

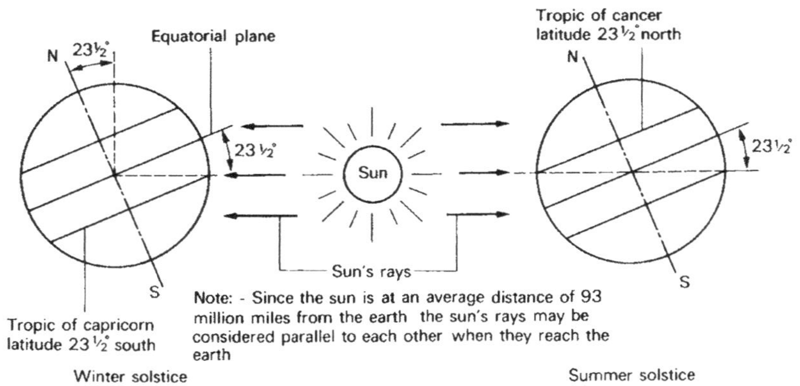

Fig. 11.1 Winter and summer solstice

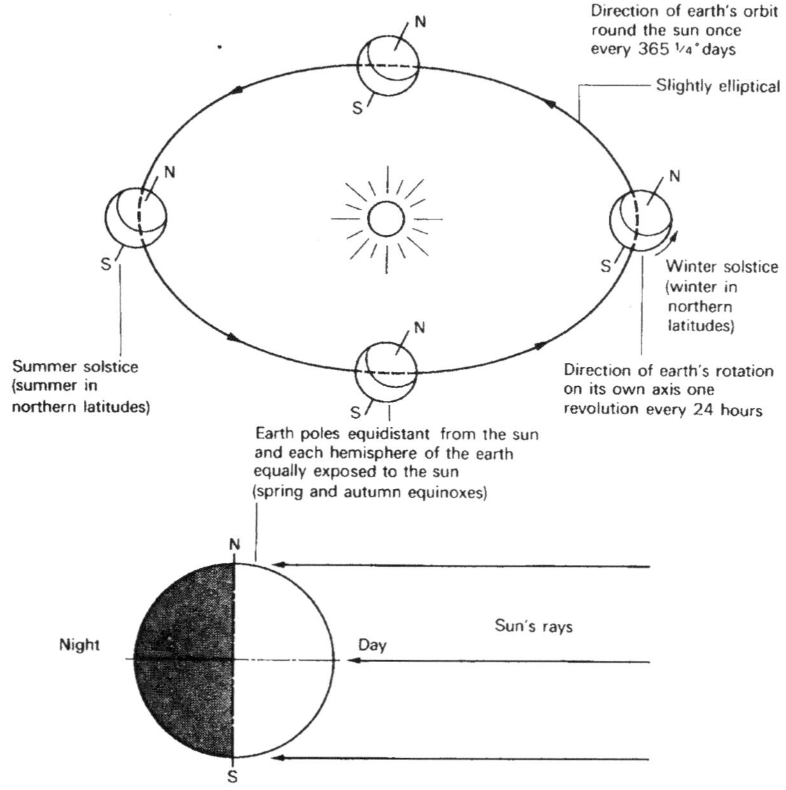

Fig. 11.2 Relative positions of earth and sun-spring and autumn equinoxes

possible to calculate the latitudes on the earth where the sun will appear directly overhead for the remaining months of the year (see Fig. 11.3).

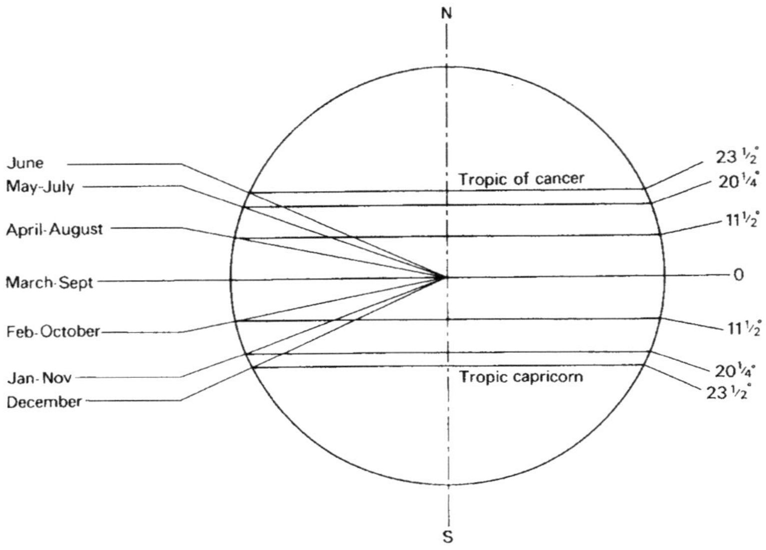

Fig. 11.3 Approximate angles of declination on the 21st of each calendar month

If required, it would be possible to calculate the relative position of the sun for every day of the year. For the purposes of solar heat gain, however, the designer is normally interested only in information for the period of maximum heat gain, which will be when the sun is at its highest position in the sky.

For latitudes lying between 23.5° north and 23.5° south, this would be when the sun is directly overhead. For latitudes outside these limits, however, the altitude of the sun must be calculated.

Example 11.1. *Calculate the maximum and minimum altitudes of the sun in Birmingham, which lies on latitude 52.5° north.*

Altitude is the angle the sun's rays make with the tangent to the earth at the altitude being considered. The greatest altitude will occur at noon on 21 June, and the least at noon on 21 December. By knowing that the declinations of the sun are 23.5° N and 23.5° S respectively, and assuming that the sun's rays are parallel, the maximum altitude can be illustrated by a diagram (see Fig. 11.4).

The following expressions for maximum and minimum altitudes can be obtained.

maximum altitude $(a) = 90 - (L - d)$

minimum altitude $(a) = 90 - [L - (-d)]$

where L = latitude being considered

d = maximum declination of the sun

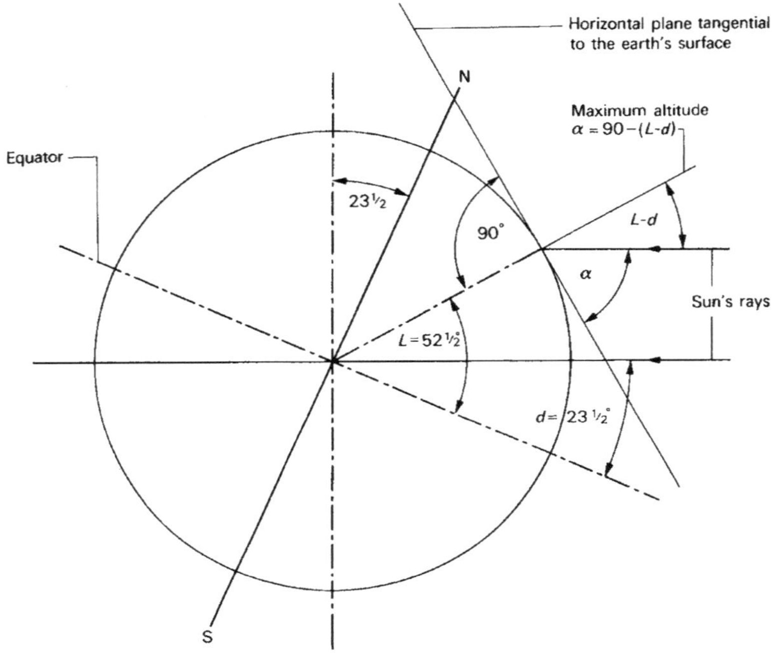

Fig. 11.4 Altitude of the sun in example 11.1

∴ the answers will be

maximum altitude $(a) = 90 - (52.5 - 23.5)$

$= 61°$

minimum altitude $(a) = 90 - [52.5 - (-23.5)]$

$\approx 14°$

Sun's position

Since the earth is rotating about its axis once every 24 hours, these altitudes will only be observed at noon; noon in this context being the time of day when the sun lies due south of the observer.

Solar azimuth angle (see Fig. 11.5)

For northern latitudes the solar azimuth angle is the angle in the horizontal plane that the horizontal component of the sun's rays make with the true north when measured from the true north in a clockwise direction. For southern latitudes, the azimuth angle is measured from the south in a clockwise direction.

Figure 11.5 shows the solar azimuth angle for northern latitudes.

It is possible, by the use of three-dimensional trigonometry, to relate the declination latitude, altitude and azimuth angles, and to calculate any of these

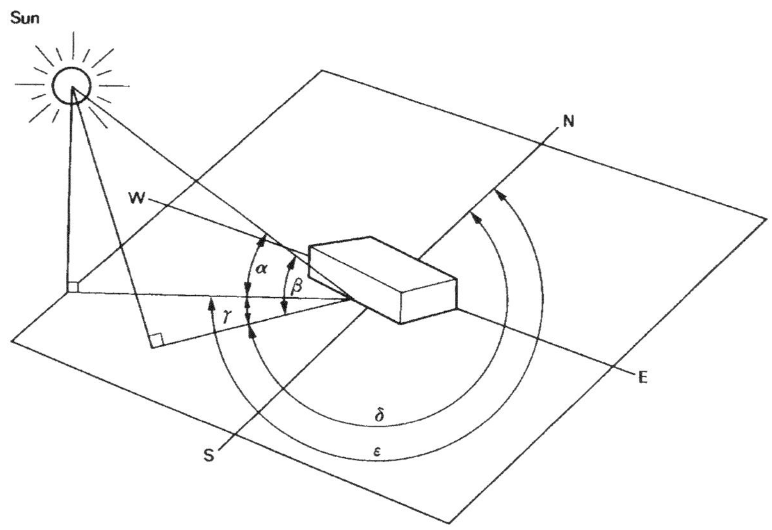

α Solar altitude angle
β Angle of incidence
γ Wall-solar azimuth angle
δ Wall azimuth angle
ε Solar azimuth angle (for northern latitudes)

Fig.11.5 Definition of angles

for any point on the earth's surface at any time. The result, obtained from such trigonometrical consideration is as follows:

$$\sin a = (\sin d \sin L) + (\cos d \cos L \cos h)$$

where
a = the altitude of the sun
d = angle of declination
L = latitude
h = hour angle

Example 11.2. *Calculate the altitude of the sun for London at 10.00 hours sun time on 21 June, given that the latitude for London is 51.5°.*

Note: Sun time is the time in hours, before or after noon; noon in this context being the time when the sun is highest in the sky.

The angle of declination, d, is 23.5° and L is 51.5° N. Since the earth revolves once every 24 hours, it will move through an angle of 360/24 = 15° in 1 hour. This angle is measured from either side of noon sun time, and therefore 10.00 hours represents an hour angle of 2 × 15° = 30°, hence:

$$\sin a = (\sin 23.5^\circ \sin 51.5^\circ) + (\cos 23.5^\circ \cos 51.5^\circ \cos 30^\circ$$

$$= (0.3987 \times 0.7826) + (0.9171 \times 0.6225 \times 0.8660)$$

$$= 0.8063$$

$$\therefore a = 53.7^\circ \text{ (approx.)}$$

If the solar azimuth angle, ε, measured east or west of south, is required, the following formula may be applied:

$$\tan \epsilon = \frac{\sin h}{(\sin L \cos h) - (\cos L \tan d)}$$

$$= \frac{\sin 30^\circ}{(\sin 51.5^\circ \cos 30^\circ) - (\cos 51.5^\circ \tan 23.5^\circ)}$$

$$= \frac{0.5}{(0.7826 \times 0.8660) - (0.6225 \times 0.4348)}$$

$$= 1.25$$

$$\therefore \epsilon = 51^\circ \text{ E or W of S (approx.)}$$

depending upon the orientation of the building.

$$\text{also } \epsilon = 180^\circ - 51^\circ = 129^\circ$$

$$\text{or } \epsilon = 180^\circ + 51^\circ = 231^\circ$$

Similar calculations may be carried out to obtain values for the altitude and solar azimuth angles for various latitudes and sun times. The Chartered Institution of Building Services (CIBS) *Guide Book A*, Section 6, provides tables of solar altitude and azimuth angles for various sun times of each month.

Sun angles

The position of the sun in relation to the orientation of any surface of a building may be specified by the angles shown in Fig. 11.5, and these include the altitude and solar azimuth angles previously described. The angle of incidence on the building face (β) is that angle between the sun's direction and the normal to the wall and is given by the following formula:

$$\cos \beta = \cos a \cos \gamma$$

where
$\cos \beta$ = the angle of incidence
$\cos a$ = the altitude of the sun
$\cos \gamma$ = the wall-solar azimuth

Intensity of solar radiation

The average intensity of solar radiation normal to the sun's rays at the outer edge of the earth's atmosphere is 1362 W/m^2 and is subject to seasonal variation of + 3.5 per cent in January and − 3.5 per cent in July. In passing through the atmosphere, part of the heat is absorbed, part scattered into space, and part scattered back to earth by the atmosphere. The portion which comes directly through the atmosphere is termed direct radiation and the portion scattered back to earth from the atmosphere is termed sky diffuse radiation. For cloudless skies, the intensities of direct and diffuse radiation will depend upon the thickness of the layer of atmosphere traversed by the sun's rays and on the solar altitude and the height above sea-level. They also depend upon the proportions of water vapour, dust and ozone in the atmosphere, which scatter and absorb radiation.

Vertical surfaces

The intensity of direct radiation on a vertical surface facing due south at noon sun time is given by the following expression:

$$I_s = I \cos a$$

where I_s = the direct incident radiation on the surface (W/m^2)

I = the intensity of direct radiation (W/m^2)

Figure 11.6 shows the intensity of direct radiation on a vertical surface facing south at noon sun time.

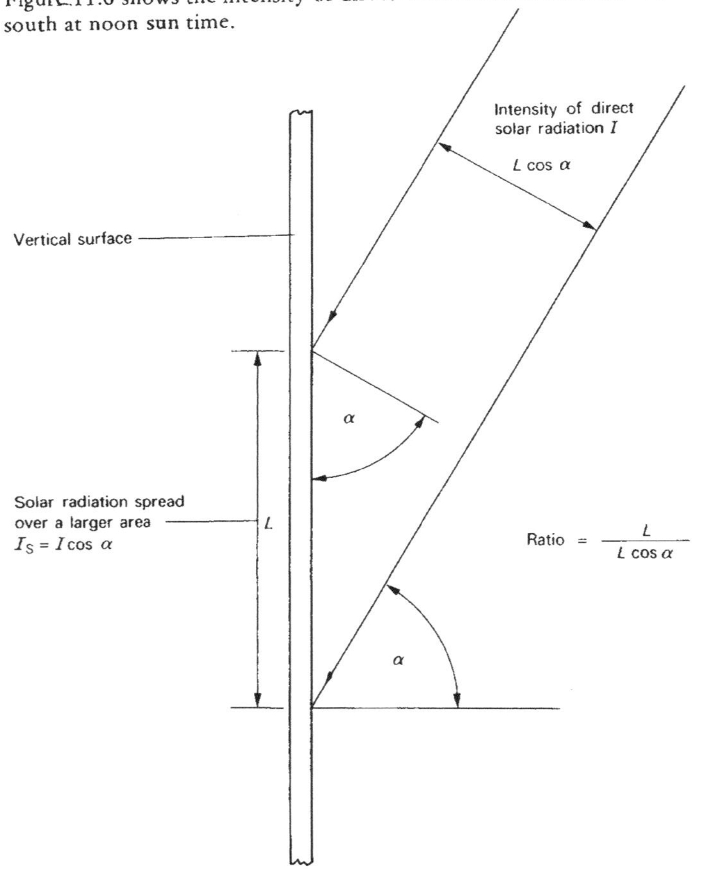

Fig. 11.6 Intensity of direct radiation on a vertical surface facing south at noon sun time

Hence, the heating effect at noon on a vertical surface facing south will be given by:

$$\text{heating effect} = \text{area} \times I \times \cos a \quad \text{(W)}$$

Example 11.3. *Calculate the total heating effect of the sun on a vertical window which faces south given the following data:*

time of day	*noon*
window size	*3 m × 2 m*
latitude	*56¼°N*
day and month	*21 July*
intensity of direct radiation normal to sun	*620 W/m²*

By reference to Fig. 11.3, on 21 July, the declination of the sun is 20¼°.

$$\text{altitude of the sun (a)} = 90 - (L - d)$$
$$= 90 - (56\tfrac{1}{4} - 20\tfrac{1}{4})$$
$$= 54^\circ$$

$$\text{heating effect} = \text{area} \times I \times \cos a$$
$$= 3 \times 2 \times 620 \times \cos 54$$
$$= 6 \times 620 \times 0.5878$$
$$= 2186.561 \text{ W}$$

Note: The intensity of direct radiation normal to the sun may be obtained from the CIBS *Guide Book A* or from the Meteorological Office.

Example 11.4. *Calculate the total heating effect of the sun in London on a vertical window which faces south given the following data:*

time of day	*noon*
window size	*2 m × 1.5 m*
latitude	*51.7° N*
day and month	*21 June*
intensity of direct radiation normal to sun	*850 W/m²*

Declination of the sun on 21 June is 23½°.

$$\text{altitude of the sun (a)} = 90 - (L - d)$$
$$= 90 - (51.7 - 23.5)$$
$$= 61.8^\circ$$

$$\text{heating effect} = \text{area} \times I \times \cos a$$
$$= 2 \times 1.5 \times 850 \times 0.4726$$
$$= 1205.13 \text{ W}$$

Horizontal surfaces

Similar calculations may be carried out to determine the heating effect on horizontal surfaces, and the following formula may be used:

$$\text{heat load} = \text{area} \times I \times \sin a$$

Example 11.5. *Calculate the total heating effect of the sun in London on a horizontal skylight using the same values given in example 11.4.*

$$\begin{aligned}\text{heat load} &= \text{area} \times I \times \sin a\\ &= 2 \times 1.5 \times 850 \times 0.8813\\ &= 2247.315\ \text{W}\end{aligned}$$

Note: The total heating effect on the horizontal surface has been increased by 1042.185 W. This is because solar radiation is spread over a smaller area on the horizontal surface, which increases its intensity (for the same altitude of the sun).

Sloping surfaces

Sloping surfaces such as pitched roofs provide additional problems of solar heat-gain calculations due to the introduction of another angle, i.e. roof slope. Two formulae have been derived for the calculation of the incident radiation on the surface, depending upon whether the roof is facing into or away from the sun.

1. Roof facing into the sun:

$$I_s = (I \sin a \cos \delta) + (I \cos a \cos \gamma \sin \delta)$$

2. Roof facing away from the sun:

$$I_s = (I \sin a \cos \delta) - (I \cos a \cos \gamma \sin \delta)$$

where I_s = incident radiation (W/m^2)
δ = angle of roof slope
a = altitude of the sun
γ = wall-solar azimuth angle

Example 11.6 illustrates the application of the formulae.

Example 11.6. *Find the intensity of solar radiation on a sloping roof of a building situated at latitude 50° N. The orientation of the ridge is 19° N of W to 19° S of E and the roof slopes at 20°. The time may be taken at 14.00 hours on 21 July.*

By reference to Fig. 11.3, the declination of the sun on 21 July is 20.25°. The hour angle for noon sun time plus two hours will be 2 × 15° = 30°.

$$\begin{aligned}\therefore \sin a &= (\sin 20.25^\circ \times \sin 50^\circ) + (\cos 20.25^\circ \times \cos 50^\circ \times \cos 30^\circ)\\ &= (0.3461 \times 0.7660) + (0.9382 \times 0.6428 \times 0.8660)\\ &= 0.7874\\ \text{solar altitude (a)} &= 52^\circ \text{ (approx.)}\end{aligned}$$

The solar azimuth angle ϵ may now be calculated as follows:

$$\begin{aligned}\tan \epsilon &= \frac{\sin 30^\circ}{(\sin 50^\circ \times \cos 30^\circ) - (\cos 50^\circ \times \tan 20.25^\circ)}\\ &= \frac{0.5}{(0.7660 \times 0.8660) - (0.6428 \times 0.3689)}\\ &= 1.1731\\ \therefore \text{solar azimuth angle } \epsilon &= 49^\circ \text{ W of S (approx.)}\\ \text{or } \epsilon = 49 + 180 &= 229^\circ \text{ SW of N}\end{aligned}$$

The wall-solar azimuth angle may now be found as shown in Fig. 11.7.

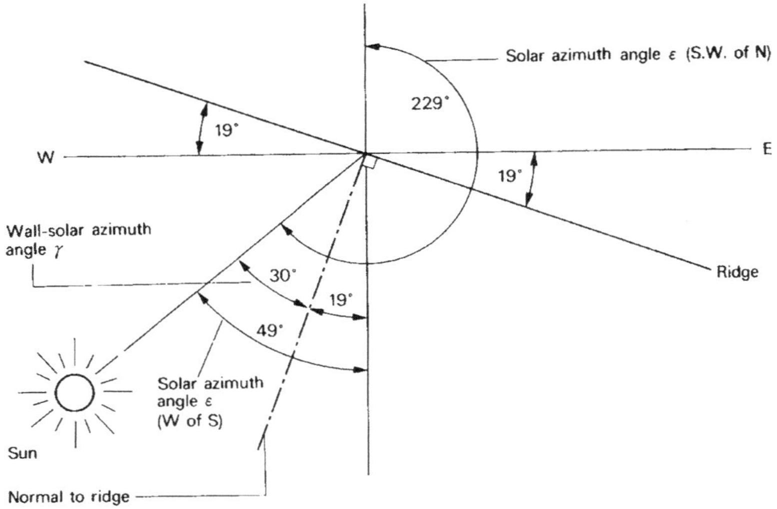

Fig.11.7 Method of finding the wall-solar azimuth angle

Since the roof is facing into the sun and inclined at 20° to the sun, the intensity of direct solar radiation normal to the sun may be taken as being about 800 W/m^2. The intensity of direct radiation on the roof on the same side of the ridge may now be found as follows:

$$\begin{aligned}I_s &= (I \sin \alpha \cos \delta) + (I \cos a \cos \gamma \sin \delta)\\ &= (800 \sin 52^\circ \cos 20^\circ) + (800 \cos 52^\circ \cos 30^\circ \sin 20^\circ)\\ &= (800 \times 0.7880 \times 0.9397) + (800 \times 0.6157 \times 0.8660 \times 0.3420)\\ &= 592.386 + 145.882\\ &= 738.3\ \text{W/m}^2 \text{ (approx.)}\end{aligned}$$

The incident radiation on the roof on the opposite side of the ridge to the sun will be as follows:

$$\begin{aligned}I_s &= (I \sin a \cos \delta) - (I \cos a \cos \gamma \sin \delta)\\ &= (800 \sin 52^\circ \cos 20^\circ) - (800 \cos 52^\circ \cos 30^\circ \sin 20^\circ)\end{aligned}$$

$= 592.386 - 145.882$

$= 446.5\ W/m^2$ (approx.)

Heat transmission through glass

The values of incident radiation so far obtained provide an estimate of the quantity of heat incident upon the outside surface of a structure. Not all this heat, however, is transmitted to the inside of the building; some is reflected or absorbed by the glass. The thermal capacity of the building fabric also has an important effect on solar heat gains. The radiation is usually incident on the floor where part is absorbed and part is diffusely reflected, to be subsequently absorbed by the walls and the ceiling.

Example 11.7. *If the roof in example 11.6 contains a skylight of single glazing measuring 1.5 m by 1 m, calculate the proportion of heat transmitted to the building using the following factors:*

glass transmission factor = *0.46*

glass absorption factor = *0.40*

percentage of absorbed radiation transmitted to the room = *68 per cent*

Roof $I_s = 738.3\ W/m^2$

heating effect on glass $= 738.3 \times 1.5\ m^2$

$= 1107.45$ W

portion transmitted to the room $= 1107.45 \times 0.46$

$= 509.43$ W

portion absorbed by the glass $= 1107.45 \times 0.4$

$= 442.98$ W

absorbed energy transmitted to the room $= 442.98 \times 0.68$

$= 301.23$ W

total heat transmitted to the room $= 509.43 + 301.23$

$= 810.66$ W

Effect of other orientations

So far, consideration of solar heat gains have been given for surfaces facing due south, i.e., buildings having an east–west line.

In order to calculate heat gains for buildings having other orientations, the wall-solar azimuth angle γ must be included, which has been described and calculated previously.

The value of the direct solar radiation upon vertical surfaces of any orientation for any time of day is found from the following formula:

$$I_s = \cos a \cos \gamma$$

where I_s = the direct incident radiation on the surface (W/m²)

I = the intensity of direct radiation (W/m²)

a = solar altitude angle

γ = wall-solar azimuth angle

Hence the heating effect on the surface will be:

$$\text{heating effect} = \text{area} \times I \cos a \times \cos \gamma \qquad (W)$$

Example 11.8. *Calculate the total heating effect of the sun on a vertical window which faces southeast given the following data:*

time of day	*noon*
altitude of sun	*58°*
latitude	*54° N*
wall-solar azimuth angle	*61° E of S*
intensity of direct radiation normal to sun	*800 W/m²*
window size	*2 m × 1.5 m*

heating effect $= \text{area} \times I \cos 58° \times \cos 61°$

$= 2 \times 1.5 \times 800 \times 0.5299 \times 0.4848$

$= 616.5$ W (approx.)

This heating effect will be reduced before it enters the room, as previously described in example 11.7.

Effect of shading

Most window frames are recessed into the wall and this causes shading of the glass and a reduction in solar heat gain into the room. A large recess is sometimes used in order to achieve this reduction in heat gain and glare.

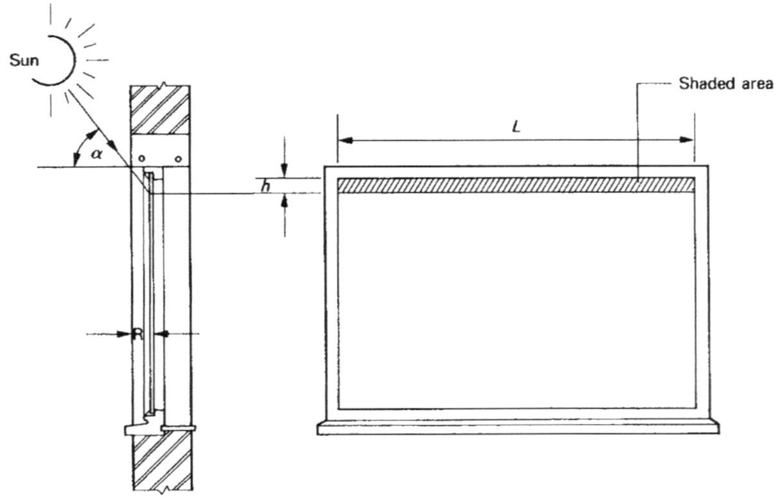

Fig.11.8 Window facing due south - depth of shade at noon sun time

For a window facing directly into the sun (which will give the highest heat gain), the shaded area can be calculated from the altitude angle of the sun (see Fig. 11.8) and the shaded area is equal to:

$$\text{shaded area} = L \times h$$
$$= L \times R \tan a$$

where L = length of window in (m)

a = altitude angle of the sun

R = depth of recess in (m)

Example 11.9. *A window frame facing directly into the sun measures 1.5 m long × 1 m high. If the altitude angle of the sun is found to be 50° and the solar radiation intensity 860 W/m², calculate the heating effect on the unshaded area of the glass when the depth of recess is 300 mm.*

$$\text{shaded area of glass} = L \times R \tan a$$
$$= 1.5 \times \tan 50^\circ \times 0.300$$
$$= 1.5 \times 1.1918 \times 0.300$$
$$= 0.5363 \text{ m}^2$$
$$\text{unshaded area of glass} = 1.5 - 0.5363$$
$$= 0.9637 \text{ m}^2$$
$$\text{heating effect on unshaded area of glass} = 0.9637 \times 860$$
$$= 828.782 \text{ W}$$

If the window is not facing directly into the sun, calculation of the shaded area involves another angle and therefore the heating effect is more difficult to determine. In Fig. 11.9 the angle γ causes the sun's rays to travel a horizontal distance y so that the depth of shading is given by:

$$h = y \times \tan a$$
$$y = \frac{R}{\cos \gamma}$$
$$\therefore h = \frac{R \tan a}{\cos \gamma}$$

The amount of side shading is given by:

$$x = R \times \tan \gamma$$

Example 11.10. *A window, not facing directly into the sun, measures 2 m long × 1.5 m high. If the altitude angle of the sun and the angle γ are found to be 38° and 36° respectively, calculate the heating effect on the unshaded area of glass when the solar radiation intensity is 750 W/m² and the depth of recess 150 mm.*

$$h = \frac{0.15 (\tan 38^\circ)}{\cos 36^\circ}$$

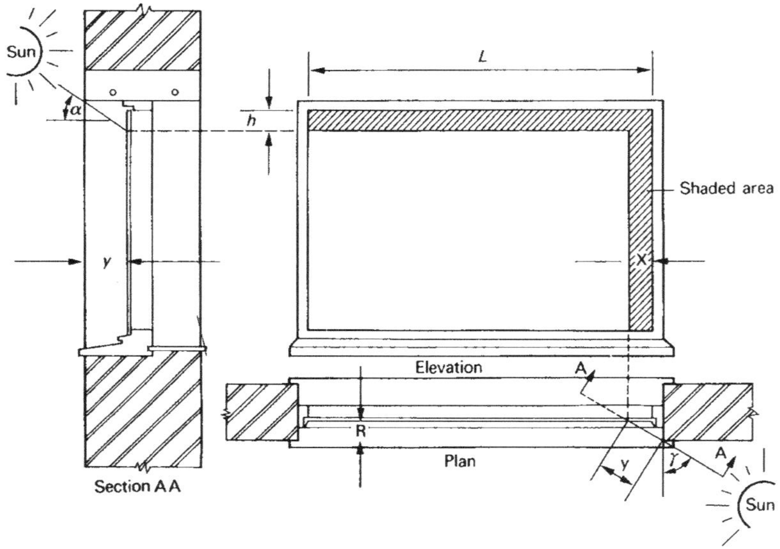

Fig. 11.9 Window in shade for any orientation

$$= \frac{0.15 \times 0.7813}{0.8090}$$
$$= 0.145 \text{ m (approx.)}$$
$$\text{and } x = 0.15 \times \tan 36^\circ$$
$$= 0.15 \times 0.7265$$
$$= 0.11 \text{ m (approx.)}$$

Therefore the total area of the window in the shade would be equal to:

$$(2 \times 0.145) + [(1.5 - 0.145) \times 0.11]$$
$$= (0.29 + 0.149)$$
$$= 0.439 \text{ m}^2$$
$$\therefore \text{unshaded area} = (2 \times 1.5) - 0.439$$
$$= 2.561 \text{ m}^2$$
$$\text{heating effect on unshaded area of glass} = 2.561 \times 750$$
$$= 1920.75 \text{ W}$$

Alternatively, the area of unshaded glass may be obtained from:

$$= (1.5 - h)(2 - x)$$
$$= (1.5 - 0.145) \times (2 - 0.11)$$
$$= 1.355 \times 1.89$$
$$= 2.57 \text{ m}^2$$

which is approximately equal to the above answer, i.e. 2.561 m^2

Note: The heating effect on the unshaded area of glass due to direct solar radiation does not include the diffused or ground-reflected radiation which has also to be considered regardless of whether the glass is in the shade.

In heavy industrial areas, where buildings are close together, the permanent smoke haze may give rise to a greater heat gain by diffused radiation than that obtained from direct solar radiation.

Heat gains through building structure

Heat gains through glass are instantaneous, whereas heat gains through the fabric of a building are delayed due to the thermal time-lag caused by the building material. Different materials will produce different thermal time-lags and heavy-weight constructed buildings have a high thermal capacity and will take longer to cool down or heat up than lightweight constructed buildings. If similar conditions exist, however, a heavyweight constructed building will require more heat input than a lightweight constructed building to bring it back to thermal comfort level.

The calculation of the maximum solar heat gain into a room therefore must include the instantaneous heat gain through the window, plus the fabric heat gain at some time previously, depending upon the thermal time-lag. To allow for this reduction in peak value, it is necessary to determine the gain to the external wall and then apply what is termed a 'decrement factor' (see Table 11.1).

Table 11.1 Adjustment factors to solar heat gains through building fabric

Construction	Adjustments time-lag (hours)	Decrement factor
Light frame (internally lined)	$\frac{1}{2}$	1.0
105 mm brickwork (internally lined)	4	0.7
220 mm brickwork (internally lined)	$8\frac{1}{2}$	0.3
150 mm concrete (internally lined)	$5\frac{1}{2}$	0.5
200 mm concrete (internally lined)	$6\frac{1}{2}$	0.4

Sol-air temperature

The calculation of the gain to the outside surface of the fabric is complicated by the fact that both the incidence solar radiation and the outside air temperature must be considered. This calculation, however, may be simplified by the concept of what is termed the 'sol-air' temperature.

The 'sol-air' temperature is defined as the theoretical outside temperature which would result in the same rate of heat transfer through the structure as exists with the actual solar radiation and the outside air temperature. The sol-air temperature at a certain time may be calculated from the following expression:

$$t_e = t_{ao} + \frac{aI}{h_{so}}$$

where
t_e = sol-air temperature
t_{ao} = actual outside air temperature
a = absorption coefficient applied to the outside surface of the building material
I = intensity of direct plus diffused solar radiation on the outside surface
h_{so} = heat transfer coefficient for the external surface

In practice, it is unnecessary to calculate the value of the sol-air temperature, as these are listed in the CIBS *Guide Book A*, Section 6.

Casual heat gains

Internal heat gains

In addition to heat gains from external sources, there will also be sensible or latent heat gains (or both) within the space itself. These gains are as follows: due to: occupants, lighting, electrical machinery, gas appliances and cooking.

Occupants: The heat gain consists of sensible heat due to radiation and convection from the body, and latent heat gain due to respiration and the evaporation of moisture from the skin. The proportion of sensible to latent heat emitted depends upon the age and sex of the occupant, the degree of activity, and the internal thermal environmental conditions.

Table 11.2 gives the total heat gain from an average adult male at different degrees of activity.

Table 11.2 Heat emission from the human body

Degree of activity	Total heat gain (W)
At rest	115
Sedentary work	140
Walking slowly	160
Light manual work	235
Medium manual work	265
Heavy manual work	440

Lighting: This comprises a sensible heat gain to the room, and the quantity of heat emitted will depend upon the type of fitting installed and the extent and usage of the fitting. Where the light fitting is specially ventilated so that the heat may be extracted, the heat transmitted to the room may be reduced below the nominal rating of the fitting. For normal light fittings 1 watt of lighting contributes 1 watt of heating.

Electrical machinery: Where electric motors are installed in the room the heat gain to the room will depend upon the following:

(*a*) the efficiency of the motor;
(*b*) whether or not the machinery being driven is also in the room;
(*c*) the frequency with which the motors will be used.

Gas appliances: The heat gain into the room from gas appliances will depend upon the following:

(*a*) the heat input to the appliance;
(*b*) the location of the appliance;
(*c*) whether it is connected to a flue or is flueless;
(*d*) position of any flue.

Cooking: The manufacturers of cooking equipment should provide details of the heat given out from their appliances. A table is given in the CIBS *Guide Book A*, Section 7.

Air infiltration (external heat gain)

If the room is maintained at a positive pressure, there will be little air infiltration, since all losses would be outwards. Where air filtrates by natural uncontrolled movement of air due to opening of doors and minute leakage through window frames, an allowance of one-half air change per hour is usually suitable.

Example 11.11. *Calculate the casual heat gains to an office given the following factors:*

1. *People 16*
2. *Lighting 20 W/m²*
3. *Floor area 80 m²*
4. *Height of room 2.5 m*
5. *Outside air temperature 30 °C*
6. *Inside air temperature 22 °C*
7. *Business machine load 400 W*
8. *Density of air 1.2 kg/m³*
9. *Specific heat capacity of air 1012 J/kg*
10. *Infiltration half air change per hour*

Internal heat gains

from people = (16 × 115) = 1840 W
from lighting = (20 × 80) = 1600 W
from machine = = 400 W
Total = 3840 W

Due to infiltration

gain = mass × specific heat capacity × temperature rise
mass = volume of room × air change × density
= 240 × 0.5 × 1.2
= 144 kg/h
= 0.04 kg/s
gain = 0.04 × 1012 × (30 − 22)
= 323.84 W

Total casual heat gains = 3840 + 323.84
= 4163.84 W

Admittance method

During warm sunny months, windows facing in a southerly direction are subject to daily cyclic heat gains from solar radiation in addition to other gains previously described.

In order to ensure that the rooms do not become uncomfortably hot during the sunny months, i.e., that the maximum peak indoor temperature does not frequently exceed about 27 °C, the CIBS *Guide Book A*, Section 8, describes a technique known as the admittance method, which enables the peak indoor environmental temperature to be assessed for any proposed building design and also gives curves of peak indoor temperature against window size for a simple design of a multi-storey building.

Routine calculations

Application of the technique requires the following data to be calculated in turn:

1. Mean heat gains from all sources.
2. Mean internal environmental temperature.
3. Swing (deviation) from mean to peak in heat gains from all sources.
4. Swing (deviation) from mean to peak internal environmental temperature.
5. From 2 and 4, the peak internal environmental temperature.

Definitions

Many of the values used for heat-loss calculations are used but some new terms are also involved, namely:

1. **Admittance factor** (Y) The factor which gives its name to the procedure, which is the amount of energy entering the surface for each degree of temperature swing at the environmental point. It is the reciprocal of the thermal resistance or impedance of an element to cyclic heat flow from the environmental temperature point and has the same units as U value (W/m² °C).

For slabs less than 75 mm thick, the admittance approximates to the U value of the structure, while for thicknesses above 200 mm, admittances tend to be a constant value. In comparison with lightweight materials, dense constructions have higher admittances, i.e., they absorb more energy for a given temperature swing.

For multi-layer slabs, the admittance is determined primarily by the surface layer, so that a 300 mm slab with 25 mm of insulation on the surface would respond more as a lightweight material than a heavy one. Table 11.3 gives admittance factors for some common types of construction.

2. **Environmental temperature** Its use is essential in the admittance method and its value takes into account both the air temperature and the mean radiant temperature, i.e.,

environmental temperature = $\frac{2}{3}$ mean radiant temperature + $\frac{1}{3}$ air temperature.

Table 11.3 Admittance factors

Construction	Admittance (Y W/m^2 °C)
External walls	
Brick solid	
105 mm unplastered	4.2
105 mm with 16 mm lightweight plaster	3.1
Brick with 20 mm cavity (unventilated)	
105 mm inner and outer leaves with 16 mm dense plaster on inner leaf	4.3
Brick with 20 mm cavity (unventilated)	
105 mm inner and outer leaves with 16 mm lightweight plaster	3.3
Internal walls (partitions)	
105 mm brick with 15 mm dense plaster on both sides	3.3
Floors and ceilings	
Suspended timber floor and plasterboard ceiling	
Floor	0.1
Ceiling	0.3
150 mm cast concrete with 50 mm screed	
Floor	5.6
Ceiling	5.6
Windows	
Single glazed (unshaded)	5.6
Double glazed (unshaded)	5.6

3. **Decrement factor** The ratio of the cyclic transmittance to the steady-state U value.

4. **Mean solar heat gains** A function of the mean incident radiation intensity as read from tables in the *Guide Book*.

5. **Mean casual heat gains** The mean heat gain from casual sources such as lighting, machinery, etc., is found by multiplying the individual items by their duration and averaging over the 24-hour cycle.

6. **Peak indoor environmental temperature** The peak indoor environmental temperature is found by adding the mean-to-swing to the mean, thus:

$$t''_{ei} = t'_{ei} + \tilde{t}_{ei}$$

where t''_{ei} = peak internal environmental temperature (°C)
t_{ei} = mean internal environmental temperature (°C)
$\tilde{t}_{ei}$ = swing in internal environmental temperature (°C)

Example 11.12. *Estimate the peak indoor environmental temperature likely to occur during a sunny period in July for the office shown in Fig. 11.10. The characteristics of the office are given in Table 11.4.*

Step 1: Mean heat gains. The solar gain is found from the following formula:

Table 11.4 Characteristics of an office for example 11.12

Item	Detail
Window	Single glazing with internal blinds
Floor	Wood block on concrete
Ceiling	Plastered concrete
Partitions	Plastered brickwork
Occupancy	10 persons for 8 hours (80 W each)
Lighting	30 W/m^2 of floor, 7.00 to 9.00 and 16.00 to 18.00
Classification	Heavyweight construction

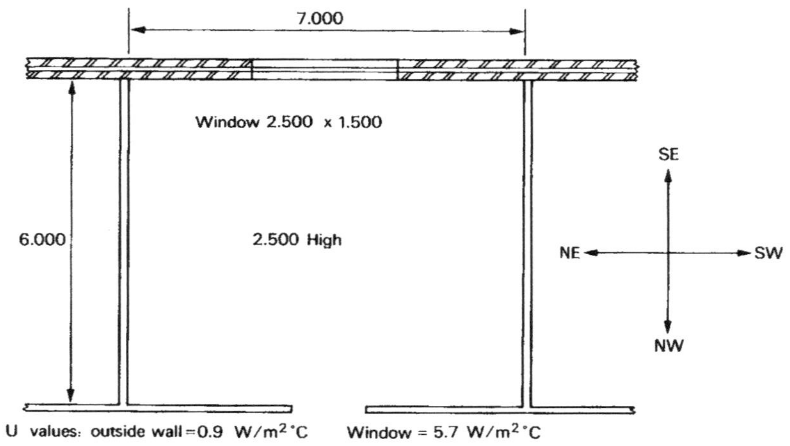

Fig. 11.10 Plan of office - example 11.12

$$Q'_s = SI'A_g$$

where Q'_s = mean solar gain (W)
I' = mean solar intensity (W/m^2)
S = solar gain factor
A_g = sunlit area of glazing

$\therefore Q'_s = 0.46 \times 190\ (2.5 \times 1.5)$
$= 327.75$ W

The mean casual heat gain is found from the following formula:

$$Q'_c = \frac{(q_{c1} \times t_1) + (q_{c2} \times t_2)}{24}$$

where Q'_c = mean casual gain (W)
q_{c1} and q_{c2} = instantaneous casual gains (W)
t_1 and t_2 = duration of individual casual gains in hours

$$Q'_c = \frac{(10 \times 80 \times 8) + (7 \times 6 \times 30 \times 4)}{24}$$

$$Q'_c = 476.7 \text{ W}$$

$$\therefore Q'_t = 327.75 + 476.7$$

$$= 804.5 \text{ W (approx.)}$$

Step 2: Mean internal environmental temperature. Assuming that the window is open during the day and closed at night, the ventilation loss is found from the following formula:

$$C_v = 0.33\, N\, v$$

where N = rate of air change per hour

v = volume of room m^3

$$\therefore C_v = 0.33 \times 3 \times 7 \times 6 \times 2.5$$

$$= 103.95 \text{ W}$$

Fabric loss

$$\Sigma AU = 5.7\,(2.5 \times 1.5) + 0.9\,(17.5 - 3.75)$$

$$= (5.7 \times 3.75) + (0.9 \times 13.75)$$

$$= 33.75 \text{ W}$$

The mean indoor environmental temperature may now be found from the following equation:

$$Q'_t = (\Sigma AU + C_v)\,(t'_{ei} - t'_{ao})$$

where ΣAU = sum of products of areas of exposed surfaces and their U values (W/° C)

C_v = ventilation loss

t'_{ei} = mean internal environmental temperature ° C

t'_{ao} = mean outdoor air temperature ° C

The daily mean outdoor air temperature for July may be taken as 19 °C.

$$\therefore 804.5 = (33.75 + 103.95)\,(t'_{ei} - 17)$$

$$\therefore t'_{ei} = 24.84\,°\text{C (approx.)}$$

Step 3: Swing (mean to peak) in heat gain. The solar gain may be found from the following equation:

$$\tilde{Q}_s = S_a A_g\,(I_p - I')$$

where $\tilde{Q}_s$ = swing in effective heat gain due to solar radiation (W)

S_a = alternating solar gain factor

I_p = peak intensity of solar radiation (W/m^2)

Peak hour is 15.00 but allow for 2-hour time-lag, so peak intensity is at 13.00.

$$\tilde{Q}_s = 0.42\,(2.5 \times 1.5) \times (280 - 190)$$

$$= 141.75 \text{ W}$$

The structural gain may be found from the following equation:

$$\tilde{Q}_f = f\,A\,U\,(t_{eo} - t'_{eo})$$

where $\tilde{Q}_f$ = swing in the effective heat input due to structural gain (W)

ϕ = time-lag in hours

f = decrement factor

t_{eo} = sol-air temperature at time of peak hours less time-lag (°C)

t'_{eo} = mean sol-air temperature (°C)

Assuming a time-lag of 6½ hours and a decrement factor of 0.4 (see Table 11.1):

$$\tilde{Q}_f = 0.4 \times 0.9\,(17.5 - 3.75) \times (28.1 - 23.7)$$

$$= 4.95 \times 4.4$$

$$= 21.78 \text{ W}$$

(This amount is small enough to be ignored.)

Casual gain $\tilde{Q}_c = Q_c - Q'_c$

$$= (10 \times 80) - 476.7$$

$$= 323.3 \text{ W}$$

Gain air to air $\tilde{Q}_a = (\Sigma A_g U_g + C_v)\,\tilde{t}_{ao}$

where $\tilde{Q}_a$ = swing in effective heat input due to swing in outside temperature in (W)

$\Sigma A_g U_g$ = sum of products of areas of exposed glazing and their U values (W/°C)

$\tilde{t}_{ao}$ = swing in outside air temperature (°C)

$$\therefore \tilde{Q}_a = [(2.5 \times 1.5 \times 5.7) + 103.95]\,6.5$$

$$= 814.6 \text{ W}$$

$$\therefore \tilde{Q}_t = 141.75 + 21.78 + 323.3 + 814.6$$

$$= 1301.43 \text{ W}$$

Step 4: Swing (mean-to-peak) in indoor environmental temperature using Table 11.3.

Floor	AY = 42	× 5.6	=	235.2	W/°C
Ceiling	AY = 42	× 5.6	=	235.2	W/°C
Window	AY = 3.75	× 5.6	=	21.0	W/°C
Outside wall	AY = 13.75	× 4.3	=	59.125	W/°C
Partitions	AY = 47.5	× 4.5	=	213.75	W/°C
		ΣAY	=	764.275	W/°C

$$\tilde{Q}_t = (\Sigma AY + C_v)\,\tilde{t}_{ei}$$

where ΣAY = sum of products of all room surface, internal and external and their appropriate admittance values (W/°C)

$\tilde{t}_{ei}$ = swing in internal environmental temperature (°C)

$$\therefore 1301.43 = (764.275 + 103.95)\,\tilde{t}_{ei}$$

$$1301.43 = 868.225 \times \tilde{t}_{ei}$$

$$\therefore \tilde{t}_{ei} = 1.5\ ^\circ C$$

Step 5: Peak internal environmental temperature.

$$t''_{ei} = t'_{ei} + \tilde{t}_{ei}$$
$$= 24.84 + 1.5$$
$$= 26.34\ ^\circ C$$

Reasonable thermal conditions should therefore exist during the period of peak internal environmental temperature.

Solar heating in the UK

Solar collectors may be sited on the roof of a building and used to reduce the cost of providing hot water, or for space heating. They may also be sited on the ground close to swimming pools and used to heat the water.

A correctly installed solar water-heating system with the collectors facing south or west of south will intercept about 1000 kWh (3.6 GJ) of solar energy each year for each square metre of net panel area.

The overall efficiency of the system taken over the year is about 40 per cent and therefore the expected useful heat output from the system will be about 400 kWh/m^2 (1.44 GJ) per year.

For the average house with a system with a 5 m^2 of net panel area, about 2000 kWh (7.2 GJ) per year may be supplied to the hot-water system and this represents about 50 per cent of the total energy used annually for water heating.

On-peak electricity costs about 2.8p per kWh and the fuel-saving in this example will be about £50 per year. If the water is heated by off-peak electricity or gas, the fuel-saving will be between £20 and £40 per year because these fuels are cheaper than on-peak electricity. If 7 GJ of energy in the form of hot water could be provided for each 19 million households in the UK, the saving in primary energy would be about 4 per cent of the total consumption.

A normal domestic installation will require about 1 m^2 for every 50 litres of daily hot-water demand.

The Building Research Establishment have carried out a computer study of solar water heating and the following points summarise their findings:

1. The possible effects of a double-glazed panel is marginal and may, for small collector areas, be negative, the improved heat-retention properties of two sheets of glass not compensating for the poorer transmission of the incoming radiation.
2. Major changes in the angle of inclination of south-facing collectors produce only small variations in annual heat output. More heat is collected in winter when the panel is inclined at 60° to the horizontal. Collectors inclined at 40° to the horizontal provide the optimum yearly heat output.
3. Collectors facing west of south are preferable to those facing east of south: due south is not necessarily the optimum direction of solar collectors.
4. For a system with a 4 m^2 of net collection area, the optimum capacity of the solar tank is 200 litres.

Figure 11.11 shows a detail of a single-glazed solar collector and Fig. 11.12 a detail of solar water-heating system which operates as follows:

1. The collector panels are heated by solar radiation which in turn heats the water inside them to a temperature of 50–60 °C. On a good summer's day it is possible for all the hot water required for an average house to be met by a system with a 4–5 m^2 collector area.
2. The pump is switched on by the control box when the temperature of the water at X exceeds that of Y by between 2–3 °C.
3. Hot water from the solar collector is pumped through the heat exchanger inside the solar cylinder which heats the water in the cylinder.
4. When the water is drawn off through the hot-water taps, the cold water from the cold-water storage cistern forces the hot water from the solar cylinder into the conventional cylinder and thus reduces or eliminates the heat required to raise the temperature of the water in the conventional cylinder to 60–70 °C. The water in the solar collector system has anti-freeze added to the water to prevent freezing. The expansion vessel takes up the expansion of the water in the solar collector system.

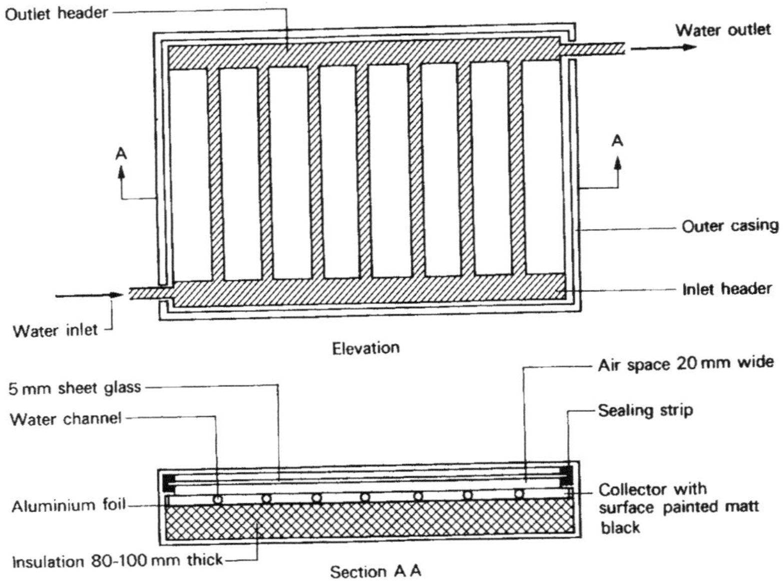

Fig. 11.11 Detail of solar collector

Solar space heating (see Fig. 11.13)

A solar-heated, timber-framed house having a floor area of 84 m^2 is planned to be built by the BRE. Incorporated into the roof will be 20 m^2 of water-heating panels which will supply heat either directly or via a heat pump, depending on the collection temperature, to a 35 m^3 insulated underground storage tank. Heat from this tank will be transferred during the night, via small heat pumps, to smaller heat accumulators of 1 m^3 and 0.3 m^3 which will supply the daily space- and water-heating requirements of the house.

The heat pumps will use electricity during the off-peak period at a reduced tariff.

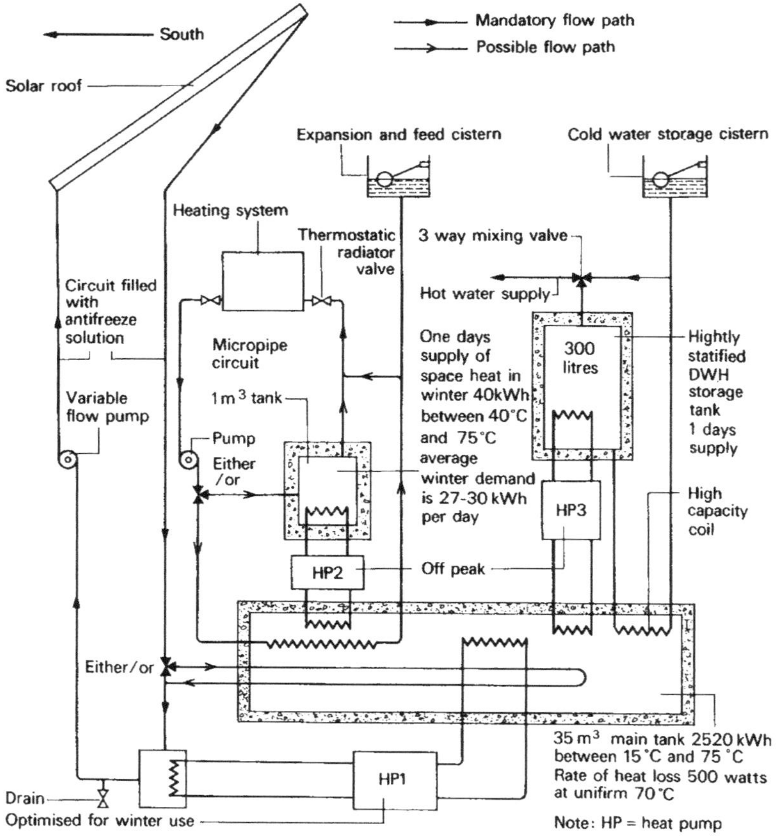

Fig.11.13 Solar heated house (BRE)

Solar heating for swimming pools

One of the more viable applications of solar-energy utilisation in the UK is the heating of swimming pools, especially those that are only in use during the summer months.

Solar collectors having no glazing or insulation will generally be satisfactory for heating swimming pools because of the small temperature rises that will be required. The collectors should be located in a sheltered but not an over-shadowed position. An area of panel between half and three-quarters of the surface area of the pool should give satisfactory results, especially if the surface of the pool is covered when not in use.

In most cases the pool water is circulated through the panels and this gives the most efficient results. Collectors should be made of material suitable for use with chlorinated water and both copper and some black plastics should be satisfactory.

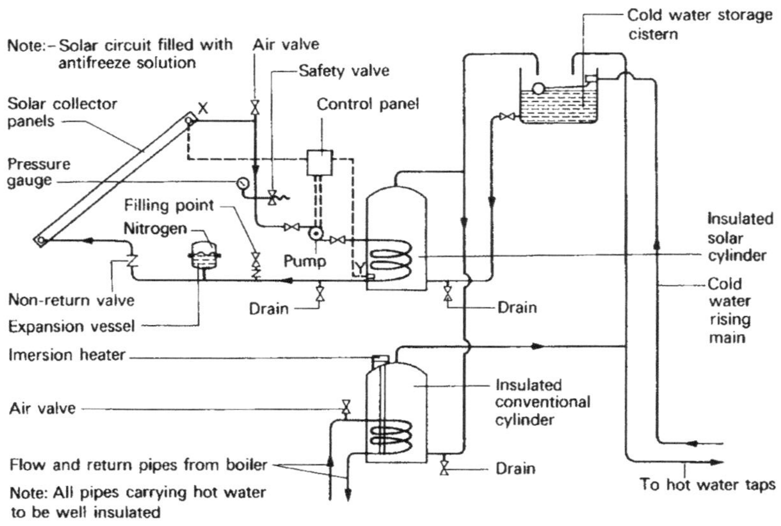

Fig.11.12 Solar water-heating system

Principles of operation of solar collector

A flat-plate collector consists of one or more cover plates of glass or transparent plastic. These materials transmit about 90 per cent of short-wave solar radiation, but allow less than 10 per cent of long-wave radiation to escape. This is often termed as the 'greenhouse effect' and ensures that solar heat is collected and not lost again by outward radiation. Coated-steel glass panels are claimed to increase the transmission of solar energy.

Questions

1. In respect to the determination of solar heat gains in buildings, define the terms altitude, azimuth, and declination.

2. Calculate the maximum and minimum altitudes of the sun in Liverpool, which lies on latitude 53.4° N.

Answers: maximum = 60.1°; minimum = 13.1°

3. Calculate the altitude of the sun for London at 16.00 hours sun time on 21 June, given that the latitude for London is 51.5° N.

Answer: 36.5° (approx.)

4. Calculate the total heating effect of the sun on a vertical window which faces south, given the following data:

time of day	noon
window size	2 m x 1.5 m
latitude	55.4° N

day and month	21 July
intensity of direct radiation normal to sun	800 W/m^2

Answer: 1380.48 W

5. Calculate the total heating effect of the sun on a horizontal flat roof given the following data:

time of day	noon
size of roof	8 m × 6 m
latitude	51.5° N
day and month	21 June
intensity of direct radiation normal to sun	850 W/m^2

Answer: 36 022.32 W

6. A horizontal glass skylight measures 3 m × 2 m. From the following data calculate the heat transmitted to the room by direct solar radiation.

time of day	noon
latitude	53.4° N
day and month	21 June
intensity of direct radiation normal to sun	760 W/m^2
glass transmission factor	0.43
glass absorption factor	0.42
percentage of absorbed radiation transmitted to room	66 per cent

Answer: 2795.56 W (approx.)

7. Calculate the total heating effect of the sun on a vertical window which faces southwest given the following data:

time of day	08.00 hours
altitude of sun	36°
latitude	51.7° N
wall-solar azimuth angle	62° W of S
intensity of direct radiation normal to sun	750 W/m^2
window size	2.5 m × 1.5 m

Answer: 1068.26 W (approx.)

8. A window facing directly into the sun measures 2 m long × 1 m high. If the altitude angle of the sun is found to be 40.5° and the solar radiation intensity 800 W/m^2, calculate the heating effect of the unshaded area of glass when the depth of recess is 150 mm.

Answer: 1395.2 W (approx.)

9. Define the term sol-air temperature, and describe how its value may be used to find the heat gains through a building structure.
10. State the sources of heat gains in a room other than those from solar radiation.
11. Explain the principle in which a flat-plate solar collector operates, and describe briefly the use of solar energy for hot-water supply and space heating.

Chapter 12

The heat pump, degree days

The heat pump (see Fig. 12.1)

The heat pump is a device which extracts thermal energy from a low-temperature source and upgrades it to a higher temperature so that it may be used for space- or water heating. The principle of operation is similar to a refrigerator, but instead of wasting the heat given out in the condenser, this heat is utilised in the heat pump.

The low-temperature source in the system may be from air, soil, or water, which surrounds the evaporator. The main advantage of the heat pump is that it always provides more energy for heating than the energy used for driving it. The device may have a reversing valve so that the evaporator (which extracts heat) and the condenser (which gives out heat) may be interchanged. With a reversing valve, the heat pump may be used for heating in winter and cooling in summer.

Coefficient of performance (COP)

The theoretical maximum coefficient of performance can be expressed as follows:

$$\text{COP (max)} = \frac{t_c}{t_c - t_e}$$

where t_c = condenser temperature in degrees Kelvin

t_e = evaporator temperature in degrees Kelvin

It will be clear from the above equation that, in common with all vapour

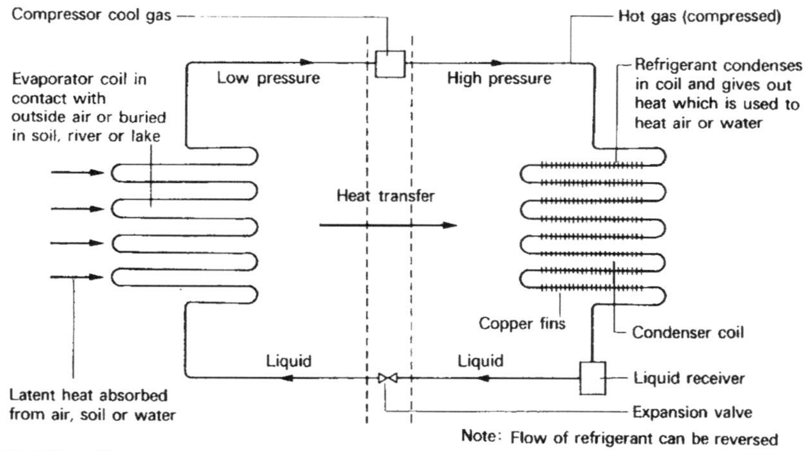

Fig.12.1 The heat pump

refrigeration systems, the heat pump operates to a greater advantage when working with a low condenser temperature and a higher evaporation temperature.

Example 12.1. *Calculate the theoretical coefficient of performance of a heat pump when the evaporator and condenser temperatures are $-1\,^{\circ}C$ and $50\,^{\circ}C$ respectively.*

$$\text{condenser temperature (absolute)} = 50 + 273 = 323\ \text{K}$$
$$\text{evaporator temperature (absolute)} = -1 + 273 = 272\ \text{K}$$

$$\text{COP (max)} = \frac{t_c}{t_c - t_e} = \frac{323}{323 - 272} = 6.33$$

If the evaporator temperature is increased, the COP will also be increased; therefore in the example 12.1, if the evaporator temperature is increased to 10 °C the theoretical COP will be:

$$\text{evaporator temperature (absolute)} = 10 + 273 = 283\ \text{K}$$

$$\therefore \text{COP (max)} = \frac{323}{323 - 283} = 8\ \text{(approx.)}$$

Theoretically, this would mean that for every 1 kW used to operate the compressor, the amount of heat received at the condenser would be 6.33 kW or 8 kW. In practice this does not happen since:

1. The actual working temperature difference between the evaporator and condenser temperatures is modified, due to the heat losses from the pipes.
2. The ideal vapour-pressure cycle requires stages of constant temperature condensation and expansion which cannot be achieved in practice.
3. The power absorbed by the fan to force the air over the condenser surfaces represents about 10 per cent of the total energy required to sustain the process.
4. Frictional losses in the compressor account for 10–15 per cent of the power required to operate the compressor.
5. The COP falls as the outside temperature falls. In practice, a realistic COP averaged over the UK heating season would be approximately 2.5, so that for every 1 kW of power consumed, 2.5 kW of useful energy would be available.

Solar collector combined with the heat pump

As mentioned earlier, the efficiency of the heat pump falls as the ambient temperature of the evaporator falls; therefore, if the temperature of air supplied to the evaporator is increased, the heat pump gives a corresponding increase in efficiency. This can be achieved by incorporating a solar collector in the system.

The solar collector consists of an air duct under the roof cladding; if single glazing is used instead of the normal roof cladding, the efficiency is increased, but at a higher capital cost.

An average increase in ambient temperature of the evaporator of 2.5 °C during the heating season is possible, which increases the heat-pump efficiency by up to 10 per cent. Figure 12.2 shows the method of using a heat pump with a solar collector.

Actual coefficient of performance

If the actual COP of the heat pump can be kept above 2.8 it is more economical to burn primary fuels at the power station to generate electricity to run the heat pump, than to burn the primary fuel directly in the building. The actual COP is defined by:

$$\text{COP}_{act} = \frac{\text{condenser heat (W)}}{\text{absorbed compressor power + heat losses (W)}}$$

Degree days

The generally accepted definition of a degree day is the daily difference in temperature in degrees Celsius between a base temperature of 15.5 °C and the 24-hour mean outside temperature (when the outside temperature falls below 15.5 °C).

For most buildings, no heating will be required when the outside temperature is 15.5 °C and due to internal heat gains from lighting, people, etc., the internal air temperature on average will be about 3 °C above the external air temperature.

Uses of degree days

The degree day may be totalled for a month and these totals used to compare the monthly changes in the weather factor, or be added together for the heating

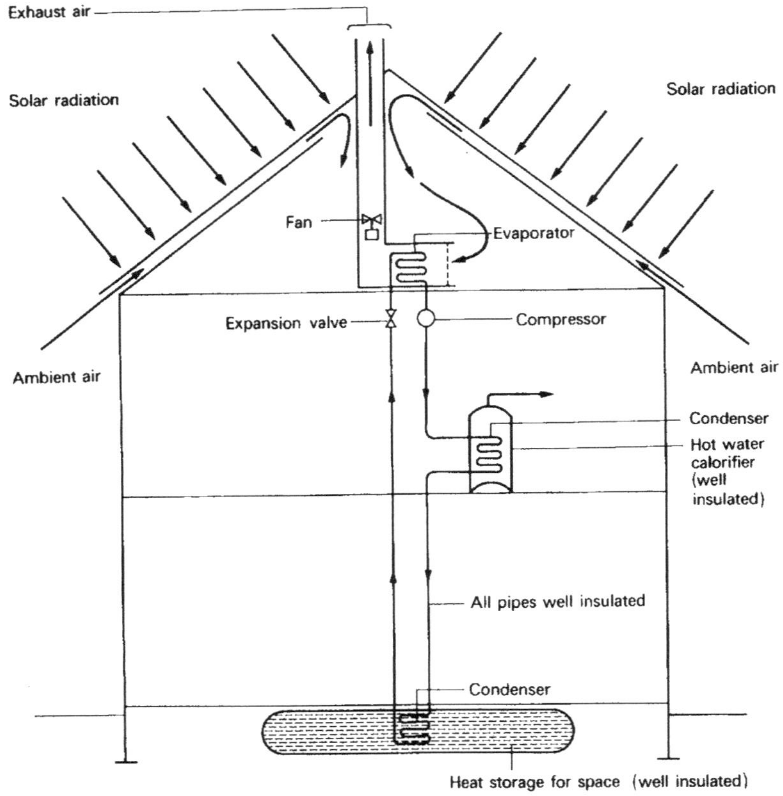

Fig.12.2 Heat pump combined with solar heating

season and thus enable the severity and duration of the winter to be compared from year to year and from place to place. Table 12.1 gives a list of monthly and yearly degree days for several areas: a full list may be obtained from the Meteorological Office.

The monthly totals may be used to check the efficiency of a heating system, by comparing the fuel used during each month, during the heating season, with the fuel used over a previous month having the same number of degree days.

Example 12.2. *Calculate the degree day over 24 hours when the external maximum and minimum temperatures are 10 °C and 5 °C respectively.*

When the daily maximum and minimum external temperatures are both below 15.5 °C the accumulated temperature in degree days may be found from the following formula:

$$\text{degree day} = 15.5 - \frac{(t_{max} + t_{min})}{2}$$

Table 12.1 Degree days for 12 months (base temperature 15.5 °C)

	Sep	Oct	Nov	Dec	Jan	Feb	Mar	Apl	May	Jun	Jul	Aug	Total
Thames Valley	51	120	262	397	373	252	238	221	134	81	25	24	2178
South Eastern	77	139	296	421	387	272	261	258	163	102	42	42	2460
Southern	81	138	268	402	370	252	245	242	153	103	36	41	2331
South Western	67	127	218	334	322	232	224	228	139	86	21	34	2032
Severn Valley	72	151	263	411	377	254	244	227	152	100	27	33	2310
Midland	100	175	302	437	414	309	281	258	182	122	48	57	2685
West Pennines	89	153	284	417	395	300	262	254	156	101	39	40	2490
North Western	99	162	291	441	419	321	271	257	185	106	48	63	2663
Borders	114	170	296	418	412	334	295	262	224	149	77	72	2823
North Eastern	99	153	322	438	417	324	280	249	184	118	53	54	2705
East Pennines	88	158	304	425	414	318	278	245	166	106	47	44	2593
East Anglia	66	142	281	422	403	289	267	250	164	105	42	42	2473
West Scotland	95	186	309	456	429	344	284	258	187	103	49	65	2765
East Scotland	108	189	319	434	430	356	300	266	221	219	67	80	2899
North East Scotland	120	200	330	455	425	380	317	287	216	148	74	91	3043
Wales	81	169	242	368	362	273	257	255	172	120	41	47	2387
Northern Ireland	100	197	301	416	409	326	278	262	190	115	45	65	2704

$$= 15.5 - \frac{(10 + 5)}{2}$$

$$= 8$$

Example 12.3. *Calculate the degree day over 24 hours when the external maximum and minimum temperatures are 18 °C and 8 °C respectively.*

When the daily maximum temperature is above 15.5 °C but by a lesser extent than the daily minimum temperature is below 15.5 °C, the accumulated temperature in degree days may be found from the following formula:

$$\text{degree day} = \frac{1}{2}(15.5 - t_{min}) - \frac{1}{4}(t_{max} - 15.5)$$

$$= \frac{1}{2}(15.5 - 8) - \frac{1}{4}(18 - 15.5)$$

$$= 3.025$$

Example 12.4. *Calculate the degree day over 24 hours when the external maximum and minimum temperatures are 20 °C and 14 °C respectively.*

When the daily maximum external temperature is above 15.5 °C but by a greater amount than the daily minimum temperature is below 15.5 °C, the accumulated temperature in degree days may be found from the following formula:

$$\text{degree day} = \frac{1}{4}(15.5 - t_{min})$$

$$= \frac{1}{4}(15.5 - 14)$$

$$= 0.375$$

Example 12.5. *A building uses 3000 kg of oil during a winter month having 380 degree days. During a previous month having the same number of degree days, the same building used 2900 kg of oil. Calculate the loss in efficiency of the heating system when compared to the previous month.*

$$\text{efficiency compared to the previous month} = \frac{2900}{3000} \times \frac{100}{1} = 96.66 \text{ per cent}$$

$$\text{loss in efficiency} = 100 - 96.66 = 3.34 \text{ per cent}$$

Example 12.6. *A building uses 5000 kg of oil during a winter month having 340 degree days. During a previous month having 320 degree days, the same building used 5200 kg of oil. Calculate the loss in efficiency.*

$$\text{oil used during current month per degree day} = \frac{5000}{340} = 14.7 \text{ kg}$$

$$\text{oil used during previous month per degree day} = \frac{5200}{320} = 16.25 \text{ kg}$$

$$\text{increase in oil used per degree day} = 16.25 - 14.7 = 1.55 \text{ kg}$$

$$\therefore \text{loss in efficiency} = \frac{1.55}{14.7} \times \frac{100}{1} = 10.5 \text{ per cent}$$

Note: Fuel consumption, however, is not solely related to the external temperature. Other factors, such as wind strength, humidity, solar radiation, cloud, or doors and windows left open, may be very significant. The use of degree days for comparisons of fuel consumed should not therefore be regarded as leading to highly accurate results.

The present official base temperature of 15.5 °C has been quoted and used in the calculations; it would, however, be reasonable to assume that the future base temperature might be 'rounded off' to 15 °C or 16 °C.

District heating

Degree days may also be used as a basis of comparison of the power used for district heating schemes in various parts of the country.

Example 12.7. *A district heating scheme of 700 dwellings in the South West required a boiler power of 9.4 MW. Calculate the boiler power required for a similar scheme in the North East. (Use figures from Table 12.1.)*

$$\text{degree days per annum in South West} = 2032$$

$$\text{degree days per annum in North East} = 2705$$

$$\text{boiler power required for North East} = \frac{2705}{2032} \times \frac{9.4}{1} = 12.5 \text{ MW (approx.)}$$

$$\text{difference} = 12.5 - 9.4 = 3.1 \text{ MW}$$

$$\text{percentage increase} = \frac{3.1}{9.4} \times \frac{100}{1} = 33 \text{ per cent (approx.)}$$

It would, if fuel costs are the same, cost about 33 per cent more for heating and hot-water supply in the North East than the South West. Degree days, however, vary from year to year and a more accurate comparison could be obtained by using the average degree days over, say, 10 years.

Questions

1. Draw a sketch of the heat pump and describe its operation.
2. Define the theoretical and actual coefficient of performance of the heat pump.
3. Calculate the theoretical coefficient of performance of a heat pump when the evaporator and condenser temperatures are 15 °C and 70 °C respectively.

Answer: 6.236

4. Define the term 'degree days' and explain how monthly totals of degree days may be used to check the efficiency of a heating system.
5. Calculate the degree day over 24 hours when the external maximum and minimum temperatures are 8 °C and 2 °C.

Answer: 3

6. A building uses 2000 kg of oil during a winter month having 410 degree days. During a previous month having the same number of degree days, the same building used 2200 kg of oil. Calculate the increase in efficiency of the heating system when compared with the previous month.

Answer: 9.1 per cent

7. State the other factors, besides external temperatures, that must be taken into account when comparing the efficiency of a heating system.
8. Explain how degree days may be used to compare the weather factors from year to year, and from place to place.

Chapter 13

Cold- and hot-water storage, expansion of materials, Boyle's and Charles' laws

Cold-water storage

Legal requirements

The minimum amount of cold-water storage depends upon the Local Water Authority Regulations. They stipulate that the minimum actual capacities of cold-water cisterns are:

1. 114 litres if the cistern is used for a direct system where all the cold-water draw-offs are supplied directly from the main. The cistern is known as a 'feed' cistern because it feeds only the hot-water cylinder.
2. 227 litres if the cistern is used for an indirect system where only the drinking water is supplied directly from the main and all other fittings are supplied 'indirectly' from what is now termed a cold-water storage cistern.

Advantages of storage

The provision of a storage cistern gives the following advantages:

1. It provides a reserve of water which can reduce the effects of a mains failure.
2. It reduces the demand on the water main.
3. Pressure fluctuations on the main are reduced.
4. It reduces the pressure on the distributing pipes which supply WC, basins, baths, showers, etc.
5. Hot-water cylinders can be substantially lighter in construction than they would be if they had to resist the mains pressure.
6. Because a storage system reduces pressure fluctuations, thermostatic mixing valves for showers, etc., operate more efficiently.
7. The reduction of pressure reduces the risk of noise from water hammer and there is less wear on the valves.
8. The air gap between the ballvalve and the water level in the storage cistern prevents water being drawn back into the main, due to 'back siphonage'.

Disadvantages of storage

1. The installation cost is increased.
2. A storage cistern should be covered with a dust-proof lid, but this is frequently omitted and as a result the water can become dirty or even contaminated by fungus growths, or mice and birds.
3. Space, structural support, and insulation must be provided for the cistern.

Foreign practice: The major difference between British and foreign practice lies in the provision of cold-water storage. Outside Britain it is very common for all water to be drawn directly from the main.

Provision of storage

For larger buildings, the capacity of the cold-water storage cistern will have to be estimated, and Table 13.1 gives the storage requirements for various types of buildings to cover 24-hour interruption of supply.

Table 13.1 Provision of cold-water storage to cover 24-hour interruption of supply. BS 6700 1987

Type of building	Storage (L)	Type of building	Storage (L)
Hostel.	90 per bed space.	Boarding school.	90 per pupil
Hotel.	200 per bed space.	Children's home.	135 per bed space
Offices		Residential nursery.	135 per bed space
with canteens.	45 per employee	Nurse's home.	120 per bed space
without canteens.	40 per employee	Nursing home.	135 per bed space
Restaurant.	7 per meal.	Convalescent home.	135 per bed space
Day schools			
nursery.	15 per pupil.		
primary.	15 per pupil.		
secondary.	20 per pupil.		
technical.	20 per pupil.		

Note The usual storage for small houses for only cold water fittings is between 100 and 150 litre, and double this amount for both hot and cold fittings.

Example 13.1. *Calculate the capacity of the cold-water storage cisterns for a five-storey office block which will have a canteen. The population will be 80 persons on each floor and a 10-hour storage of water in case of interruption of supply has been decided.*

$$\text{storage capacity for 24-hr interruption of supply} = 5 \times 80 \times 45 = 18\,000 \text{ litres}$$

$$\text{storage capacity for 10-hr interruption of supply} = \frac{18\,000}{1} \times \frac{10}{24} = 7500 \text{ litres}$$

It is not always possible at the early design stage to know the number of people that will occupy a building, but the number and types of sanitary fittings will be known.

Table 13.2 may be used to find the storage requirements.

Table 13.2 Capacities of hot and cold water added together, required for single use of appliances

Appliance	Capacity in litres
Wash basin	
hand wash	5
hand and face wash	10
hair wash	20
Shower	40
Bath	110
WC	10
Washing machine	150
Sink	
wash-up	15
cleaning	10

Example 13.2. *Calculate the capacity of a cold-water storage cistern for a dwelling containing bath, WC, wash basin, and sink.*

During the peak demand period it is usual for a dwelling to allow consecutive use of a bath and single use of other fittings. The wash basin will normally be in the bathroom so this may be ignored; also the cold water to the sink will be supplied directly from the main so a smaller amount of cold water will be required to feed the hot-water cylinder.

$$
\begin{aligned}
2 \text{ baths} \times 110 &= 220 \\
1 \text{ WC} \times 10 &= 10 \\
1 \text{ sink} \times 5 &= 5 \\
\hline
&\quad 235 \text{ litres}
\end{aligned}
$$

This is above the minimum capacity of 227 litres required by the regulations. If the mains supply is good, the water entering the cistern during the time that water is drawn off should permit a 227-litre storage being used satisfactorily.

The above method may also be used to estimate the minimum cold-water storage for larger buildings. The following steps may be taken:

1. Prepare a worksheet as shown in Table 13.3 and enter all types of sanitary appliances with their requirements obtained from Table 13.2.
2. Check the hourly pattern of use of sanitary appliances with the building owner and estimate the highest possible demand. For each hour of the day, multiply the number of times each type of appliance is likely to be used, by its capacity, and enter the product under the hour of day.
3. When all the sanitary fittings have been completed, add them under their respective columns and enter each total in the gross-demand row.
4. Hourly rates of water flow into the cistern should be obtained from the Local Water Authority and entered in the recovery-rate row. The figures

Table 13.3 Estimation of water consumption

Types of appliances and consumption (litres)	Hours of day 1	2	3	4	5	6	7	8	9	10	11	12	13	14	15	16	17	18	19	20	21	22	23	24
Basins						30	50	100	100	50	50	100	50	50	50	50	150	100	100	50	100	100	100	50
Baths							550	550	550	330	330	330	330	330	220	220	330	550	550	550	220	220	220	220
Showers						80	320	320	200	160	160	160	120	200	200	80	80	320	320	320	160	160	80	80
WCs						100	200	200	100	100	50	50	200	50	100	50	50	200	200	200	200	200	200	100
Sinks						25	100	100	100	50	50	50	100	100	100	50	50	50	100	100	100	50	50	
Gross Demand						235	1220	1270	1050	690	640	690	800	730	670	450	660	1220	1270	1220	780	730	650	450
Recovery rate						700	700	700	700	700	700	700	700	700	700	700	700	700	700	700	700	700	700	700
Net loss or gain						+ 465	− 520	− 575	− 350	+ 10	+ 60	+ 10	− 100	− 30	+ 30	+ 250	+ 40	− 520	− 570	− 520	− 80	− 30	+ 50	+ 250

t_c = temperature of cooler water in °C
t = final temperature of mixture in °C

By transposition

$$q_1 (t_h - t) = q_2 (t - t_c)$$
$$(q_1 \times t_h) - (q_1 \times t) = (q_2 \times t) - (q_2 \times t_c)$$
$$(q_1 \times t_h) + (q_2 \times t_c) = (q_1 \times t) + (q_2 \times t)$$
$$(q_1 \times t_h) + (q_2 \times t_c) = t(q_1 + q_2)$$
$$\therefore t = \frac{(q_1 \times t_h) + (q_2 \times t_c)}{q_1 + q_2}$$

Example 13.7. *A tank containing 50 litres of water at 70°C has 20 litres of cold water at 10°C mixed with it. Determine the resultant temperature of the whole. Ignore heat losses.*

$$t = \frac{(50 \times 70) + (20 \times 10)}{50 + 20}$$
$$t = 52.86\,°C \text{ (approx.)}$$

Example 13.8. *A three-way thermostatic mixing valve allows 5 litres at 60°C to mix with 8 litres at 40°C. Determine the resultant temperature of the water discharging from the valve.*

$$t = \frac{(5 \times 60) + (8 \times 40)}{5 + 8}$$
$$= 47.69\,°C$$

In ventilating and air-conditioning systems, recirculated air is often mixed with fresh air. The temperature of the resultant mixture may be found by use of the same equation.

Example 13.9. *In an air-conditioning system 15 kg of fresh air at 8°C is mixed with 10 kg of recirculated air at 22°C. Determine the resultant temperature of the mixture.*

$$t = \frac{(10 \times 22) + (15 \times 8)}{10 + 15}$$
$$= 13.6\,°C$$

Example 13.10. *A recirculating fan hot-water convector heater is used to heat a room, the heat losses from which are 5 kW. If the convector handles 0.08 m^3/s of air to provide a room air temperature of 22°C, calculate the temperature of the air leaving the convector.*

heat capacity of air (HC) = 1.34 kJ/m^3K

$$kW = m^3/s \times HC \times (t_1 - t_2)$$

where t_1 = temperature of the air leaving the convector (°C)
t_2 = temperature of the air in the room (°C)

By transposition

$$t_1 = \frac{kW}{m^3/s \times HC} + t_2$$
$$= \frac{5}{0.08 \times 1.34} + t_2$$
$$= 68.64\,°C$$
$$= 69\,°C \text{ (approx.)}$$

Expansion of materials

Various materials used in services do not expand equally when their temperatures are raised. In order to determine the increase in length of a material due to an increase of temperature, the coefficient of linear expansion of the material must be known.

The coefficient of linear expansion is defined as the fraction of its original length by which a substance expands per degree rise in temperature. Table 13.7 gives the coefficient of linear expansion of some common materials used in services.

Table 13.7 Coefficient of linear expansion of common metals and plastics used in services

Aluminium	
cast	0.000 025 5
sheet	0.000 023 0
Brass	
cast	0.000 018 7
sheet	0.000 019 3
Copper	
sheet	0.000 017 4
tube	0.000 016 9
Iron	
cast	0.000 010 2
Lead	
pipe and sheet	0.000 029 3
Steel	
mild	0.000 011 3
nickel chrome	0.000 015 3
invar	0.000 001 0
Tin	0.000 021 4
Zinc	
sheet	0.000 026 1
Polyethylene	
low density	0.000 28
high density	0.000 11
Polyvinyl chloride	
normal impact	0.000 05
high impact	0.000 08

Example 13.11. *Calculate the linear expansion of 15 m of copper tube when heated from 10 °C to 70 °C.*

expansion = coefficient of linear expansion × temperature rise × length

expansion = 0.000 016 9 × (70 − 60) × 15

= 0.015 21 m or 15.21 mm

Example 13.12. *A straight length of mild-steel heating pipe 200 m long will be heated from 15 °C to 80 °C. Expansion joints that will each accommodate 76 mm of the expansion are to be inserted in the straight length. Calculate the number of expansion joints that will be required.*

expansion = 0.000 011 3 × (80 − 15) × 200

= 0.1469 m or 146.9 mm

$$\text{number of joints} = \frac{146.9}{76} = 1.933$$

Therefore two expansion joints would be required.

Expansion of water

The expansion of water on being heated may be found from the following expression:

$$E = V\left[\frac{\rho_1 - \rho_2}{\rho_2}\right]$$

where E = expansion of water in litres or m^3

V = volume of water before being heated in litres or m^3

ρ_1 = density of water before being heated in litres or m^3

ρ_2 = density of water after being heated in litres or m^3

Table 13.8 gives the density of water at various temperatures.

Table 13.8 Density of water at various temperatures

Water temperature (°C)	Water density (kg/m^3)	Water temperature (°C)	Water density (kg/m^3)
0	999.80	60	983.00
4	1 000.00	63	982.00
10	999.70	65	980.70
15.6	990.00	66	980.10
20	998.00	70	977.50
25	996.50	71	977.00
30	995.00	74	975.00
38	993.00	80	972.00
40	992.00	82	970.40
46	989.00	91	965.00
50	987.50	100	958.00

Example 13.13. *A low-temperature hot-water heating system contains 12 m^3 of water at a temperature of 10 °C before being heated. Calculate the expansion of the water in the system when heated to a design temperature of 80 °C.*

from Table 13.8,

density of water at 10 °C = 999.70 kg/m^3

density of water at 80 °C = 972.00 kg/m^3

$$\text{expansion} = 12\left[\frac{999.70 - 972}{972}\right]$$

$$= 12 \times \frac{27.7}{972}$$

$$= 0.342 \text{ m}^3 \text{ (approx.)}$$

Feed and expansion cistern (see Fig. 13.2)

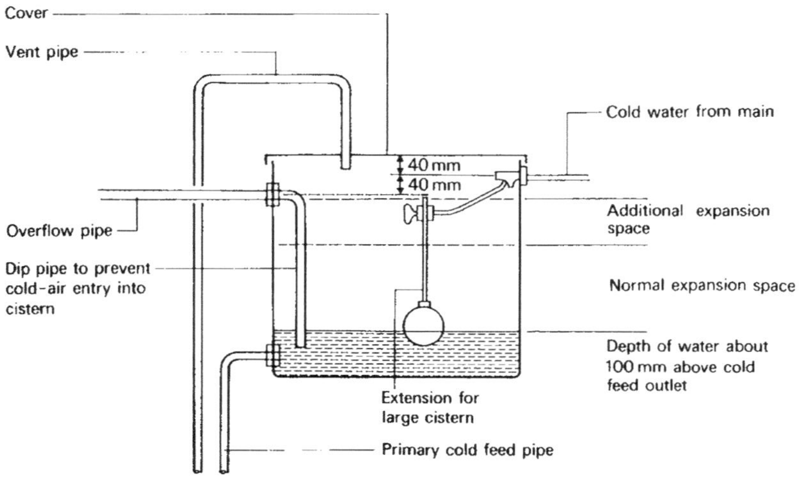

Fig. 13.2 Feed and expansion cistern

In a low-temperature heating system, a feed and expansion cistern is used to fill the system with cold water and also to accept the expanded water when the system heats up. In practice, the depth of cold water in the cistern is about 100 mm and the rest of the space is left for the expanded water, plus an additional expansion space having a capacity of not less than 4 per cent of the total capacity of water in the primary circuit, including the boiler (Water Bylaw).

The gas laws

Boyle's law (see Fig. 13.3)

This law states that for a fixed mass of gas at constant temperature, the volume is inversely proportional to the pressure:

$$P = \frac{C}{V}$$

$$\text{or } PV = C$$

where P = absolute pressure

V = volume

C = constant

Also $P_1 V_1 = P_2 V_2$

The application of the above equation enables any difference in volume to be determined.

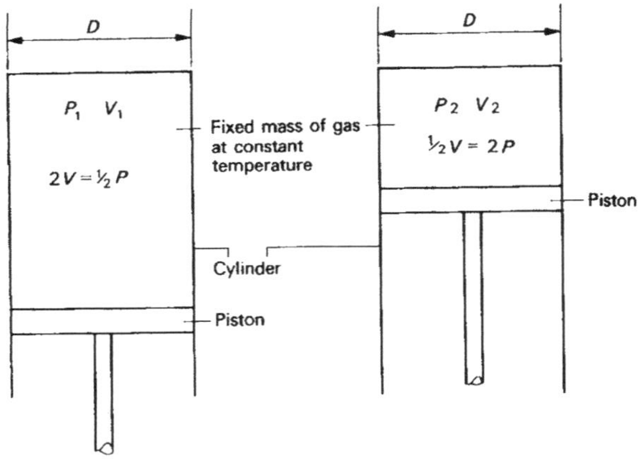

If the volume of the gas in the cylinder is halved the pressure is doubled and vice versa also $3V = \frac{1}{3}P$ and $\frac{1}{3}V = 3P$, etc

Fig. 13.3 Boyle's law

Example 13.14 (see Fig. 13.4). *The initial volume of air at atmospheric pressure in a pneumatic cylinder for boosting cold water is 3.5 m^3. When the pumps are switched on this volume is reduced to 0.8 m^3. Calculate the pressure of air in the cylinder for boosting the water. Atmospheric pressure can be taken as 101.33 kPa.*

$$P_1 = 101.33 \text{ kPa}$$
$$V_1 = 3.5 \text{ m}^3$$
$$V_2 = 0.8 \text{ m}^3$$
$$P_1 V_1 = P_2 V_2$$
$$\therefore P_2 = \frac{P_1 V_1}{V_2}$$
$$= \frac{101.33 \times 3.5}{0.8}$$

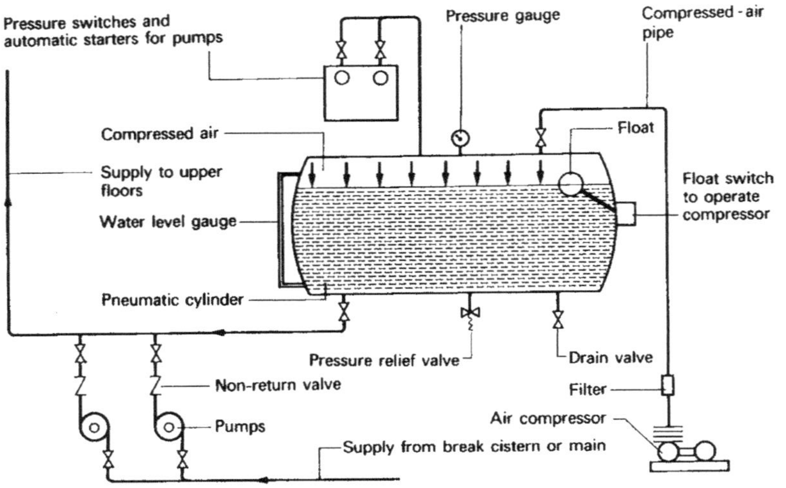

Fig. 13.4 Automatic pneumatic cylinder

$$P_2 = 441.875 \text{ kPa (absolute)}$$
$$= 441.875 - 101.33 \text{ (gauge)}$$
$$= 340.545 \text{ kPa (gauge)}$$

Theoretically, this would force the water up to a height of:

$$\text{height} = \frac{340.545}{9.81} = 34.7 \text{ m (approx.)}$$

This would result in a height of about 30 m, depending upon the frictional losses due to the pipes and fittings.

Charles' law

This law states that for a fixed mass of a perfect gas at constant pressure, the volume is directly proportional to its absolute temperature. The law is derived from the fact that a given mass of a perfect gas will change its volume 1/273 of its volume at 0 °C for a change in temperature of 1 °C, provided that the pressure remains constant.

By combining Boyle's law and Charles' law, a formula is obtained which takes into account the effect of varying pressure, volume, and temperature (see Fig. 13.5):

$$\frac{P_1 V_1}{T_1} = \frac{P_2 V_2}{T_2}$$

where P_1 = initial pressure

P_2 = new pressure

V_1 = initial volume

V_2 = new volume

T_1 = initial temperature

T_2 = new temperature

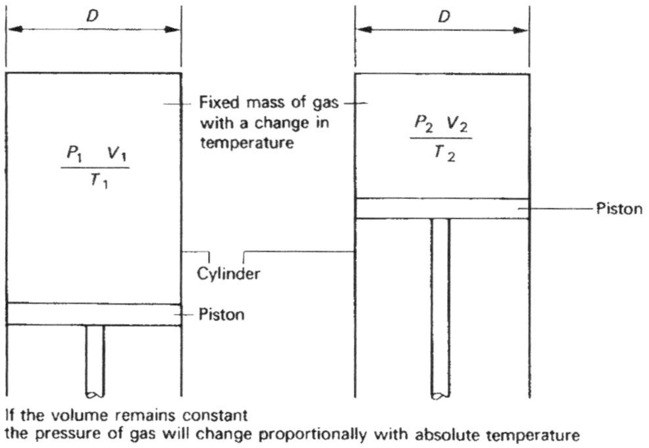

Fig.13.5 Combination of Boyle's and Charles' laws

Example 13.15. *A compressed-air cylinder has a volume of 5 m³ and contains air at atmospheric pressure and a temperature of 20 °C. If the air temperature in the cylinder is raised to 80 °C, calculate the new pressure in the cylinder.*

P_1 = 101.33 kPa (absolute)

V_2 = 5 m³

V_1 = 5 m³

T_2 = 70 + 273 = 343 K

T_1 = 20 + 273 = 293 K

$$\frac{P_1 V_1}{T_1} = \frac{P_2 V_2}{T_2}$$

$$P_2 = \frac{P_1 \times V_1 \times T_2}{V_2 \times T_1}$$

$$= \frac{101.33 \times 5 \times 343}{5 \times 293}$$

= 118.62 kPa (absolute)

= 118.62 − 101.33 = 16.29 kPa (gauge)

Drainage testing

When drainage pipelines are tested by use of air, it is essential to take into account the fall in air pressure due to the air temperature. If the air that is pumped into the pipeline is higher in temperature than the air already inside the pipeline, its volume will be reduced, showing a drop in the U-gauge water level. In order to obtain a correct reading of the U-gauge, about 15 minutes should be allowed so that the air pressure stabilises.

Example 13.16 (see Fig. 13.6). *A 225 mm soil stack is subjected to an air test recorded by a water filled U-gauge. At an initial temperature of 20 °C the gauge showed a reading of 40 mm head of water. Calculate the head of water in the gauge when the air temperature inside the pipe falls to 15 °C.*

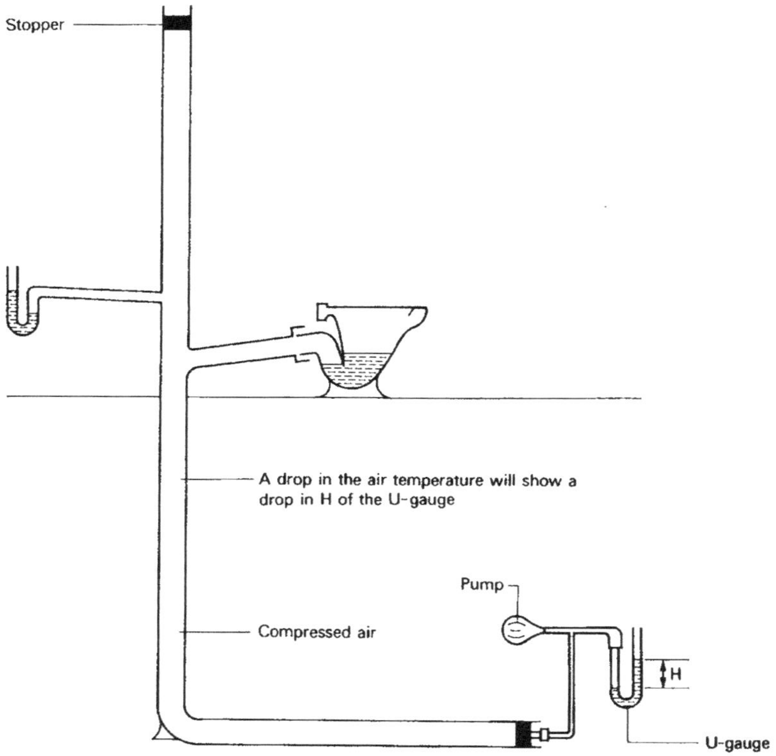

Fig.13.6 Air test on soil stack

$$\frac{P_1 V_1}{T_1} = \frac{P_2 V_2}{T_2}$$

$$\frac{40 \text{ mm} \times V_1}{20 + 273} = \frac{\text{head of water} \times V_2}{15 + 273}$$

V_1 and V_2 are equal and are therefore cancelled

$$\therefore \frac{40}{293} = \frac{\text{head of water}}{288}$$

$$\text{head of water} = \frac{40 \times 288}{293}$$

$$= 39.32 \text{ mm (approx.)}$$

Example 13.17. *5 m³ of a gas at an initial pressure of 50 kPa (gauge) and at a temperature of 30 °C is compressed in a cylinder to 3 m³, the pressure of the gas being raised to 155 kPa (gauge). Calculate the rise in temperature of the gas after compression.*

$$P_1 = 50 + 101.33 \text{ kPa (absolute)}$$

$$P_2 = 155 + 101.33 \text{ kPa (absolute)}$$

$$V_1 = 5 \text{ m}^3$$

$$V_2 = 3 \text{ m}^3$$

$$T_1 = 30 + 273 = 303 \text{ K}$$

$$\frac{P_1 V_1}{T_1} = \frac{P_2 V_2}{T_2}$$

$$T_2 = \frac{T_1 \times P_2 \times V_2}{P_1 \times V_1}$$

$$= \frac{303 \times 256.33 \times 3}{151.33 \times 5}$$

$$= 308 \text{ K (approx)}$$

$$\text{rise in temperature} = 308 - 303$$

$$= 5°\text{C}$$

Example 13.18. *16 m³ of air at 10 °C and at a pressure of 180 Pa (gauge) is passed through a heater battery. If the temperature of the air is raised to 50 °C and the pressure falls to 80 Pa (gauge), calculate the change in volume of the air.*

$$P_1 = 101.33 + 0.180$$

$$P_2 = 101.33 + 0.08$$

$$V_1 = 16 \text{ m}^3$$

$$T_2 = 50 + 273 = 323 \text{ K}$$

$$T_1 = 10 + 273 = 283 \text{ K}$$

$$\frac{P_1 V_1}{T_1} = \frac{P_2 V_2}{T_2}$$

$$V_2 = \frac{P_1 \times V_1 \times T_2}{T_1 \times P_2}$$

$$= \frac{101.51 \times 16 \times 323}{283 \times 101.41}$$

$$= 18.279 \text{ m}^3$$

$$\text{change in volume} = 18.279 - 16$$

$$= 2.279 \text{ m}^3$$

Questions

1. State the Water Authorities' legal requirements for the minimum storage of cold water in buildings and explain the merits and demerits of cold-water storage.

2. Calculate the capacity of the cold-water storage cisterns required for a hotel having 200 residents, including staff. A 24-hour storage is required in case of interruption of supply.

Answer: 27 200 litres

3. A five-storey office block is to be constructed having the following sanitary fittings on each floor: 6 WCs, 8 wash basins, and 1 sink. Prepare a worksheet of the probable amount of water used during the time between 07.00 and 18.00 hours and from the values draw a graph to find the minimum cold-water storage requirements. The water supply from the main is calculated to supply 9000 litres per hour for refilling the cistern.

4. Calculate the hot-water storage requirements for a general hospital having 600 patients and staff.

Answer: 16 200 litres

5. If the hot-water calorifiers for the hospital in question 4 are to be duplicated for maintenance or repair, calculate the power in kW of each boiler when each calorifier will have a capacity of $\frac{2}{3}$ of the total storage. Use the following factors:

   (*a*) Temperature rise of water = 55 °C
   (*b*) heating-up time or recovery period = 2 hours
   (*c*) efficiency of plant = 60 per cent
   (*d*) specific heat capacity of water = 4.2 kJ/kg K

Answer: 57.75 kW (or 58 kW approx.)

6. Estimate the hot-water storage for a sports pavilion having the following sanitary fittings: 10 showers, 8 WCs, 8 wash basins, and 2 cleaning sinks.

7. A tank containing 150 litres of cold water at 10 °C has 100 litres of hot water at 80 °C mixed with it. Determine the resultant temperature of the mixture. Ignore heat losses.

Answer: 38 °C

8. Calculate the linear expansion of 60 m of steel tube when heated from 20 °C to 80 °C.

Answer: 46.8 mm

9. A low-temperature hot-water heating system contains 45 m³ of water at a temperature of 15.6 °C before being heated. Calculate the expansion of the water in the system when heated to a design temperature of 74 °C.

Answer: 0.6923 m³

10. Define Boyle's law and state how it may be applied to various problems in services when the temperature remains constant.

11. The initial volume of air at atmospheric pressure in an air cylinder used to absorb the shock due to water-hammer in a cold-water system is 0.3 m^3. When the water is turned on, the volume of air is reduced to 0.05 m^3. Calculate the pressure of water in kPa on the water main.

Answer: 607.98 kPa (absolute) and 506.65 kPa (gauge)

12. Define Charles' law and state how it may be applied to problems in services when the temperature is changed.

13. A compressed-air cylinder has a volume of 6 m^3 and contains air at atmospheric pressure and at a temperature of 20 °C. If the air temperature in the cylinder is raised to 100 °C, calculate the new pressure in the cylinder.

Answer: 130 kPa (absolute) and 28.67 kPa (gauge)

14. 20 m^3 of air at 25 °C and at a pressure of 120 Pa (gauge) is passed through a cooling coil. If the temperature of the air is lowered to 10 °C and the pressure raised to 150 Pa (gauge), calculate the change in volume of the air.

Answer: $V^2 = 18.988$ m^3 and change in volume = -1.012 m^3

Chapter 14

Passenger lifts, electrc lighting

Passenger lifts

The total capacity of passenger lifts required for any building, to give a certain grade of service, is determined by the number of occupants and the number of visitors expected to use the lifts. The information may be obtained from the building owner but if this is not possible, an assessment of the probable building population will have to be made based on the net floor area. Surveys have shown that the space standards may vary from 1 person per 4 m^2 to 1 person per 20 m^2, but for overall population assessment, experience from existing buildings shows that 1 person per 9.5 m^2 to 1 person per 11.25 m^2 gives the average range.

Transportation during peak period

Surveys of existing buildings show that between 10 per cent and 25 per cent of the total building population will require transportation during a 5-minute peak period, according to whether they start or finish work at different times.

If no information is available, for speculative development or where staggered starting and finishing times are practised, 12 per cent of the building population per 5 minutes may be assumed.

For single-purpose buildings, or where unified starting and finishing times are practised, an average peak requirement of 17 per cent of the building population per 5 minutes may be assumed.

Number of lifts

The number of lifts will have an effect on the quality of service provided. Four

12-person lifts will provide the same carrying capacity as three 16-person lifts, but the waiting time will be about twice as long with the latter group of lifts. It is therefore essential to assess the quality of service provided along with the considerations of space required and capital cost.

Interval for lifts

The interval is expressed in seconds and represents the round-trip time of one car divided by the number of cars in a common group or system. It provides a criterion for measuring the quality of service. The average waiting time may therefore be expressed theoretically, as half this interval, but in practice it is probably nearer three-quarters of the interval.

Intervals may be assessed as in Table 14.1.

Table 14.1 Intervals for a group of lifts

Interval (seconds)	Quality of service
25–35	excellent
35–45	acceptable for offices
60	acceptable for hotels
90	acceptable for flats

The minimum number of lifts required are given in Table 14.2.

Table 14.2 Minimum number of lifts

Installation	Quality of service
1 lift for 3 floors	excellent
1 lift for 4 floors	average
1 lift for 5 floors	below average

Calculation of lift performance

The calculation of lift performance depends upon the following:

1. acceleration and retardation times;
2. car speed;
3. speed of door operation;
4. performance with variations of car load.

The probable upwards peak performance and hence the round-trip time of a lift car may be calculated from consideration of the following:

1. total passenger-transfer time;
2. total door-operating time;
3. total running time.

Running time

When a lift departs from the lift lobby at the peak period it will stop at several floors on its upward journey. The number of stops may be estimated from the following probability formula:

$$S_1 = S - S\left(\frac{S-1}{S}\right)^n$$

where S_1 = the probable number of stops
S = the maximum number of stops
n = number of passengers

The probable number of stops may be found from Table 14.3.

Table 14.3 Probable number of stops at 80 per cent load

Contract load persons	Probable number of stops, S, for the stated maximum number of stops, S			
	5	10	15	20
6	4	4	4	4
8	4	5	5	6
10	4	6	6	7
12	5	6	7	8
16	5	7	9	10
20	5	8	10	11
24	5	9	11	13

When the probable number of stops are known, the time of the upwards journey may be calculated, taking into account the car speed and the effect of acceleration and retardation at each stop. The following formula may be used:

$$T_u = S_1\left(\frac{L}{SV} + 2V\right)$$

where T_u = total upward journey time in seconds
L = total lift travel in metres
V = car speed in m/s

For the downward journey it may be assumed that during the peak period the car will run non-stop from top to bottom and the following formula may be used:

$$T_d = \left(\frac{L}{V} + 2V\right)$$

where T_d = total downward journey time in seconds
L = total lift travel in metres
V = car speed in m/s

Door-operating time

The round-trip time for a lift car will depend largely upon the time the car is stationary at the various floor levels while the passengers are entering or leaving the car.

A car with wide and shallow proportions will permit quicker passenger transfer than a car of narrow and deep proportions. Centre-opening doors are much more efficient than side-opening doors. Passengers will move towards the centre-opening entrance as soon as the doors are open, whereas with single-panel or two-speed side-opening doors, the panel will have to move half-way across the entrance before the passengers start to move.

The time taken for door operation during the round-trip time is therefore dependent upon the width of the opening, the type of door, and the door-operating speed. The time may be found from the following formula:

$$T_o = 2\,(S_1 + 1)\,\frac{W}{V_d}$$

where T_o = door operating time (opening and closing) in seconds
S_1 = the probable number of stops
W = width of opening in m
V_d = door-operating speed

For centre-opening doors, an operating speed of 0.4 m/s may be taken, and if two-speed doors are used this time is reduced to 0.2 m/s.

Passenger transfer time

For a car with wide, shallow proportions and centre-opening doors, the average total time taken for a person to enter or leave the car may be taken as 2 seconds, but this time would have to be increased to about 3 seconds for a car having narrow and deep proportions.

The total passenger-transfer time may therefore be found from the following formula:

$$T_p = 2n \text{ or } T_p = 3n$$

where T_p = total passenger transfer time in seconds
n = number of passengers

Round-trip time (RTT)

This is the time in seconds taken by a single lift car to travel from the lowest floor served to the top floor and back again. The round-trip time may therefore be calculated from the following formula:

$$\text{RTT} = T_u + T_d + T_o + T_p$$

where T_u = total upward journey time in seconds
T_d = total downward journey time in seconds
T_o = door operating time in seconds
T_p = total passenger transfer time in seconds

From the round-trip time the interval, capacity, and quality of service may be determined.

Example 14.1. *A group of four lift cars each having a carrying capacity of 20 persons and a car speed of 2.5 m/s is specified to serve a sixteen-storey hotel having room heights of 3 m each. The net floor area above ground level is to be 8000 m² and a population density of 1 person per 10 m² of net floor area is to be served. Calculate the round trip time for one lift car, the interval, and capacity for the group and the quality of service provided.*

car travel = 3×16 = 48 m

probable number of stops (Table 14.3) = 10

1. $$T_u = S_1\left(\frac{L}{SV} + 2V\right) = 10\left(\frac{48}{15 \times 2.5} + 2 \times 2.5\right) = 10\,(1.28 + 5) = 10 \times 6.28 = 62.8\text{ s}$$

2. $$T_d = \left(\frac{L}{V} + 2V\right) = \left(\frac{48}{2.5} + 2 \times 2.5\right) = 19.2 + 5 = 24.2\text{ s}$$

3. $$T_o = 2\,(S_1 + 1)\,\frac{W}{V_d}$$

Assuming a door width of 1.2 m,

$$T_o = 2\,(10 + 1)\,\frac{1.2}{0.4} = 2 \times 11 \times 3 = 66\text{ s}$$

4. $$T_p = 2n$$

Allowing for a load factor of 80 per cent,

$$T_p = 2 \times 20 \times 0.8 = 32\text{ s}$$

round-trip time = 185

$$\text{interval for the group} = \frac{185}{4} = 46\text{ s (approx)}$$

$$\text{capacity of the group} = \frac{5 \times 60 \times 4 \times 20 \times 0.8}{185}$$

= 104 persons per 5 minutes

Table 14.1 shows that an acceptable quality of service for an hotel is an interval of 60 seconds and therefore an interval of 46 seconds is satisfactory. The actual flow rate may now be found allowing a population density of 1 person per 10 m² of net floor area above ground level and that 12 per cent of this population will require transportation during the 5-minute peak period.

$$\text{actual flow rate} = \frac{8000 \times 96}{10 \times 100} = 96 \text{ persons}$$

The capacity of the group is 104 and therefore the number, speed, and capacity of the lift installation are satisfactory.

Electric lighting

The inverse-square law

The intensity of illumination from a point source varies inversely as the square of the distance from the source.

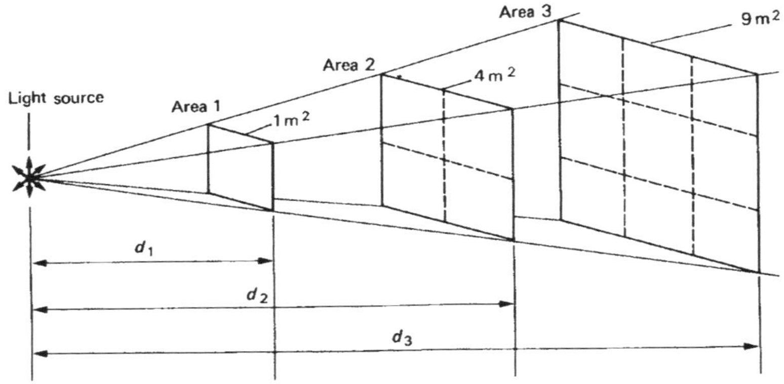

If $d_1 = 1$ m and area 1 = 1 m², then area 2 = 4 m², area 3 = 9 m²

Fig.14.1 The inverse-square law

By reference to Fig. 14.1 it will be seen that the area over which the luminous energy, from the light source, will be spread increases with the square of the distance from the source.

The illumination produced on a surface area in m² from a light source perpendicular to the surface at a distance d from the luminous intensity may be found from the following expression:

$$E = \frac{I}{d^2}$$

where E = the illumination produced on the surface in lux
I = the illumination intensity of the light source in candelas
d = the distance from the light source to the surface in metres

Example 14.2. *A spotlight having a light intensity along its beam of 16 000 candelas is directed on to a horizontal surface 4 m from the light source. If the beam of light is normal to the surface, calculate the maximum illumination level on the surface in lux.*

$$E = \frac{I}{d^2} = \frac{16\ 000}{4^2} = 1000 \text{ lux}$$

Cosine law of illumination

If the surface is not normal, i.e., perpendicular to the direction of the emitted light from the source, the illumination on the surface will be reduced because the light flux is spread over a greater area. The illumination on the surface will be less by cos θ, where θ is the angle between the light beam and the normal to the surface. This is the second basic law of illumination and is usually referred to as the cosine law of illumination.

The formula given previously will therefore have to be modified as follows:

$$E = \frac{I \cos \theta}{d^2}$$

Example 14.3. *Figure 14.2 shows a light source at* A *directed on to a horizontal surface at* B. *If the intensity of light at* A *is 10 000 candelas, calculate the illumination on the horizontal surface.*

$$E = \frac{10\ 000 \times 0.866}{5^2} = 346.4 \text{ lux}$$

The lumen (see Fig. 14.3)

The lumen may be defined as the flow of light through an area of 1 m² on the surface of a sphere of 1 m radius with a uniform point source of 1 candela at its centre. The total number of lumens emitted in all the space around this source is equal to the area in square metres in the surface of a sphere having a radius of 1 m.

The surface area of a sphere may be found from the formula $4\pi R^2$ and since the radius of the sphere is 1 m, its surface area will be 4π, or 12.57 m² approximately.

Since the luminous energy decreases with the square of the distance from the light source, the intensity of illumination may be found from the following expression:

$$E = \frac{F}{12.57 \times d^2}$$

where F = flux in lumens
d = distance from the light source to the surface

Lux: This is the unit of the level of illumination and an illumination of 1 lumen per square metre is equal to 1 lux.

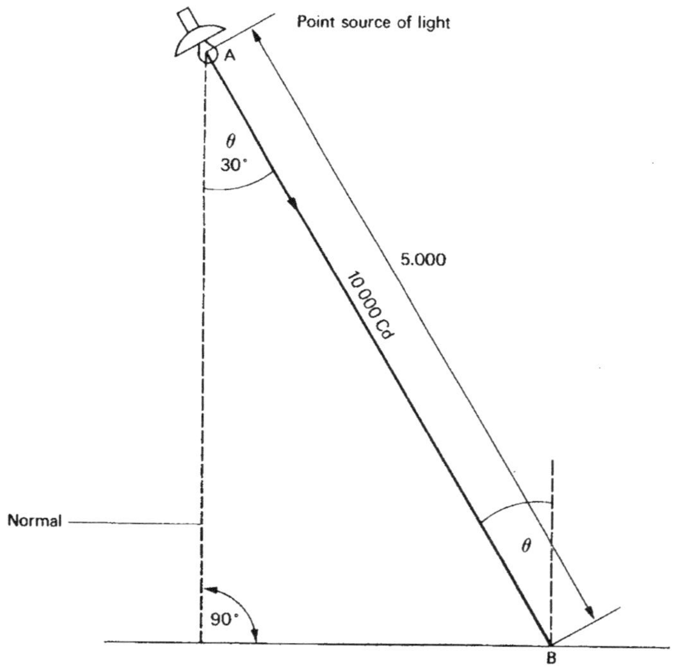

(a) Illumination on a surface when the beam of lights light is at 90° to the surface $= \frac{I}{d^2}$

(b) Illumination on a surface when the beam of light is at another angle θ to the surface $= \frac{I \cos \theta}{d^2}$

Fig.14.2 Illumination from a point source to a surface - example 14.3

Example 14.4. *A diffusing spherical light-fitting is suspended so that its centre is 2 m above the working plane. If the fitting emits 1500 lumens uniformly in all directions, calculate the intensity of illumination on the working plane.*

$$\text{intensity of illumination} = \frac{1500}{12.57 \times 2^2}$$

$$= 29.83 \text{ lux}$$

Average illuminance on a horizontal plane

The service illuminance provided on a horizontal plane throughout a building interior by means of overhead lamps depends on the number and luminous flux of the lamps used, the properties of the lamp, the dimensions of the room, the room surface reflectances, and the standard of maintenance. The illuminance is usually specified and the lighting designer, having selected suitable lamps and luminaires, will have to calculate the number of lamps required and their spacings to meet the specification.

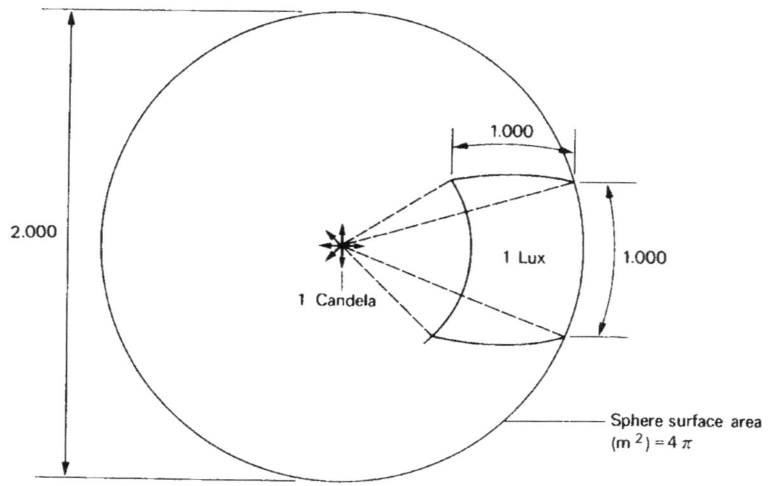

A light intensity of 1 candela on an area of the surface of the sphere of one square metre = 1 lux
total light output = 4π I lumens

Fig.14.3 The light intensity and lux

Lumen method of design

This is still the most commonly used method of lighting design and its objective is to determine a lighting layout that will provide a specified service illuminance on the horizontal working plane from an installation of luminaires mounted overhead in a substantially regular pattern.

The method requires the total luminous flux in lumens that will provide the specified illuminance to be determined, hence the term 'lumen method of design'. The following formula is used in the lumen method of design:

$$E = \frac{F \times N \times U \times M}{A}$$

where E = the average horizontal illumination at the working plane in lux

F = the lamp lighting design lumens

N = number of lamps

U = the utilisation factor

M = the maintenance factor

A = the area of the working plane in m^2

Utilisation factor

This is a factor which takes into account the performance of the light fittings, the shape of the room, and the reflectances of the room surfaces; its value must be modified by the maintenance factor.

Maintenance factor

This takes into account the light lost due to dirt on the fittings and on the room surfaces. For air-conditioned rooms a factor of 0.9 may be used. For normal conditions a factor of 0.8 may be used, while for industrial atmosphere, where cleaning is difficult, a factor as low as 0.5 may sometimes be used.

Room index

There is an infinite range of room dimensions, but it has been found that the behaviour of light in rooms is a function, not of the room dimensions, but of the room index, which is the ratio of the area of the horizontal surfaces to that of the vertical surfaces of the room.

For the lumen method of design the vertical surfaces are measured from the working plane to the centre of the fittings. This is expressed by the following equation:

$$\text{room index} = \frac{\text{length} \times \text{width}}{(\text{length} + \text{width}) \times (\text{height of fitting above the working plane})}$$

When all the values are known the lighting installation may be designed using the following steps:

1. Decide upon the illumination required in lux.
2. Calculate the room index.
3. From manufacturers' tables find the utilisation factor for the luminaire to be used.
4. Assume a suitable maintenance factor.
5. Calculate the number of fittings from the formula of the lumen method of design.
6. From Table 14.4 find the ratio of spacing to mounting height of fitting.
7. Draw the layout of the fittings to a suitable scale.

Table 14.4 Ratio of spacing to mounting height of fittings.

British Zonal (BZ) classification	Maximum spacing/mounting height ratio
1 and 2	1 : 1
3 and 4	1.25 : 1
5 to 10	1.5 : 1

Example 14.5. *A general office measuring 15 m × 9 m × 3 m high is to be illuminated to a design level of 400 lux using 85 W fluorescent fittings having a BZ classification of 3. The fittings are to be flush with the ceiling and the working plane is to be 850 mm above the floor. Design the lighting system for the office when the installed flux is 8000 lumens per fitting.*

The utilisation factor can be found from manufacturers' tables from the value of the room index.

$$\text{room index} = \frac{L \times W}{H\,(L + W)}$$

$$= \frac{15 \times 9}{1.65\,(15 + 9)}$$

$$= 3.4$$

The utilisation factor from tables is found to be 0.56 and a maintenance factor of 0.8 may be assumed.

$$E = \frac{F \times N \times U \times M}{A}$$

by transposition

$$N = \frac{E \times A}{F \times U \times M}$$

$$N = \frac{400 \times 15 \times 9}{8000 \times 0.56 \times 0.8}$$

$$N = 15$$

In terms of illumination, 15 fittings would provide 398 lux and would probably be satisfactory.

In terms of spacing, however, 16 fittings would be required which would provide the following illumination level:

$$E = \frac{F \times N \times U \times M}{A}$$

$$E = \frac{8000 \times 16 \times 0.56 \times 0.8}{15 \times 9}$$

$$E = 424.77 \text{ lux}$$

Spacing: The fittings have a BZ classification of 3 and therefore the maximum spacing to mounting height ratio is 1.25 : 1 (see Table 14.4).

mounting height = 3 − 0.85 = 2.15 m

maximum spacing = 2.15 × 1.25 = 2.7 (approx.)
(centre-to-centre of fittings)

The distances of the fittings from the wall should not exceed half of the above spacing, and less if there is a working surface near to the wall. The maximum distance from the centre of the fittings to the wall is therefore 2.7/2 = 1.35 m.

Figure 14.4 shows one method of spacing the fittings for the office.

Glare index

The glaring effects in a room can be evaluated by the glare index. A glare index of 10 indicates glare which would normally be perceptible, while a glare index of 30 indicates glare which would normally be intolerable.

The glare index for any room can be calculated from first principles, but it is time-consuming. When lighting is arranged in a regular pattern of sources in or near the ceiling, the glare index can be obtained from tables published by the Illuminating Engineering Society (IES) and the Lighting Industry Federation Ltd., or by use of a circular calculator.

Table 14.5 gives limiting glare indexes for various types of buildings.

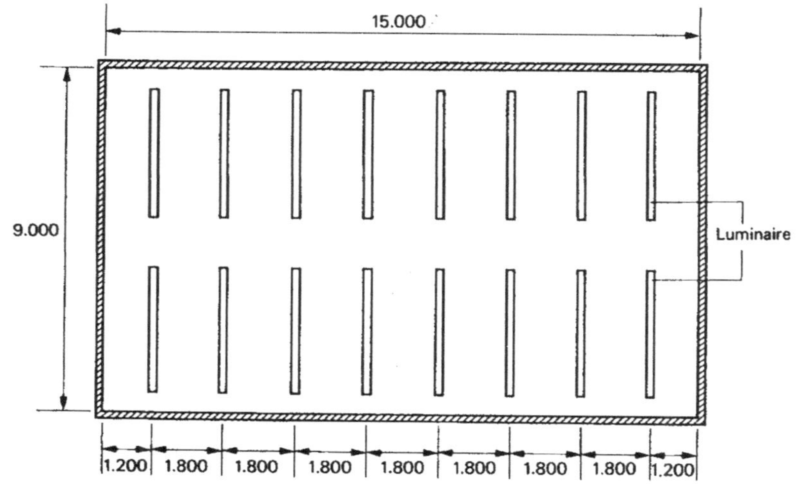

Fig. 14.4 Layout of light fitting – example 14.5

Table 14.5 Limiting glare indexes (IES)

Location	Limiting glare index
Offices – general	19
Offices – drawing	16
Schools – classrooms	16
Hospitals – wards	13
Factories – rough assembly	28
Factories – precision assembly	19

Cost of lamp replacement

To calculate the cost of replacement of lamps, the following formula may be used:

$$AC_{av} = \frac{10\,n \times N \times T \times b \times W}{L} - (n \times N \times T)$$

where AC_{av} = average annual cost of replacement
n = number of lamps per fitting
N = number of fittings
T = net cost per lamp (if starter switches are used add 50 per cent of the net cost of one starter switch)
b = number of working hours per week
W = number of working weeks in the year
L = rated life of lamps in hours

Table 14.6 gives the rated life of various types of lamps.

Table 14.6 Rated life of lamps

Rated life of lamps	
Fluorescent tubes	7 500 hours
High-pressure sodium	6 000 hours
Mercury filament	7 500 hours
Tungsten–halogen	2 000 hours
Tungsten–filament	1 000 hours (double-life type 2 000 hours)
Mercury discharge	7 500 hours
Gold–cathode discharge	20 000 hours

Example 14.6. *Calculate the annual cost of replacement for 16 fluorescent luminaires. Use the following data:*

(*a*) *cost per tube* = *£6.00*
(*b*) *cost of starter switch* = *£4.00*
(*c*) *working hours per week* = *40*
(*d*) *working weeks per year* = *25*

$$AC_{av} = \frac{10 \times 16 \times (6+2) \times 40 \times 25}{7500} - (16 \times (6+2))$$

$$= 170.67 - 128$$

$$= 42.67 \text{ or } £43 \text{ (approx.)}$$

Check list for design of electric lighting systems

The Illuminating Engineering Society Code for Interior Lighting 1977 provides the following check list for the lighting of normal building interiors. The list may have to be modified for special lighting schemes.

General requirements

1. Purpose of the interior, probable layout of plant, furniture or equipment.
2. Availability of daylight and depth of daylight penetration, the need for a combined electric/daylighting system and the automatic control of the electric lighting.
3. Visual difficulty of the task and the risk of veiling reflections.
4. Task and general surround illuminances and whether the task is on the vertical or horizontal plane; local, localised or general lighting and the need for optical aids.
5. Strength of modelling required and the provision of directional lighting.
6. Colour appearance and colour rendering required.
7. Limiting glare index and other requirements for glare content.
8. Statutory requirements or official recommendations.
9. Need for emergency lighting.

Environmental conditions and requirements

10. Presence or absence of hostile environment and the need for special luminaires, i.e., flameproof, dust-tight, or watertight.
11. Unusually high or low ambient temperatures, e.g., foundries, cold stores, and the effect on control gear and luminaire component; the effect on light output of fluorescent lamps.
12. Possibility of high ambient temperatures near luminaires, i.e., at ceiling level and shop windows.
13. Possible effect of radiant heat from furnaces or other industrial equipment on luminaires and control gear.
14. Effect of heat from luminaires on air temperature in the interior and the use of lighting heat to supply part of the building heating.

Effect of structural features

15. Dimensions of the interior; length, width, and height.
16. Luminaire mounting height.
17. Luminaire spacing/mounting ratio.
18. Reflectances of ceiling, walls, and floor including the influence of furnishings, windows and glazed partitions.
19. Co-ordination of lighting equipment with other building services.
20. Limitations of luminaire mounting position, i.e., roof structure, number of bays, modular construction of building, space available in the ceiling void.
21. Effects of obstruction by parts of the structure, i.e., beams, ventilation ducting, pipework, heavy plant and furniture.

Lamps

22. Lamp types which meet colour appearance and colour rendering requirements for the interior or activity.
23. Lamp light output related to source, size and mounting height.
24. Rationalisation of lamp type, colour and wattage, particularly with existing installations. The availability of replacements.
25. Run-up time to full light output and the need for standby system to cover interruption of power supply.
26. Need to reduce flicker and stroboscopic effect.
27. Economics; capital and running costs.

Luminaires

28. Suitability for purpose.
29. Appearance.
30. Luminous intensity distribution required.
31. Authenticated photometric data.
32. Need for special equipment.
33. Ease of maintenance.
34. Availability; present and future of spare parts.
35. Weight and fixing arrangements.
36. Use of trunking systems or lighting tracts.

Maintenance

37. Maintenance factors.
38. Accessibility of luminaires and the need for special equipment.
39. Acceptability of proposed luminaires, lamps, etc., to clients' maintenance staff.

Use of energy

40. Check of power loading for proposed scheme with suggested target loading.
41. Check of switching or control systems with the object of minimising the use of energy.

Heat from lighting

Only a small proportion of the total electrical energy consumed by a luminaire is converted into light, and all the energy used will be converted into heat. The overheating of rooms by the use of tungsten lamps can occur at relatively low levels of illumination. If 100 W tungsten lamps are used having an efficiency of about 50 per cent to provide an illumination level of 400 lux for offices, the installed lighting load would provide about 70 W/m^2 of heat.

If white fluorescent lamps are used to provide 400 lux, the installed lighting load would provide about 20 W/m^2 of heat. If deluxe fluorescent lamps are used having an efficiency of 50 per cent to provide 400 lux, the installed lighting load would increase by 50 per cent to 30 W/m^2.

At a lighting intensity of 1000 lux, up to 75 per cent of the heat that is generated inside a building is produced by the lighting, which results in a considerable cooling load on the air-conditioning plant. In order to remove this heat. from the space, ventilated lighting fittings can be used and the heat transferred by ducting to other parts of the building that require heating, i.e., perimeter of rooms. Figure 14.5 shows the energy balance for: (*a*) surface-mounted; (*b*) recessed; and (*c*) ventilated luminaires, connected to an extract duct.

Surface-mounted luminaires give all the energy to the room, and the air at ceiling level will be warmer than at the floor level; currents of cool air at floor level may result in discomfort to the occupants. Recessed luminaires disperse about 60 per cent of the generated heat into the void between the ceiling and the floor above, and this results in a reduction of the heat entering the room to 23 per cent, which still results in the air at ceiling level being warmer than the air at floor level. Ventilated luminaires extract almost the full amount of the heat generated by the fittings out of the room. The heating effect on the room therefore results in less circulation of the air and the elimination of unpleasant draughts. The cooling effect of the air passing round the tubes helps to optimise the level of lighting efficiency and also helps to keep the tubes and fittings clear of dust. Ventilated luminaires can provide an improvement of 10 to 12 per cent in lighting efficiency when compared to unventilated types.

By reducing the window areas of a building to not more than 20 per cent of the wall areas and providing a high level of electric lighting and insulation, the heat extracted by the luminaires, people and machinery is usually sufficient to heat the building even when the air temperature outside is below freezing.

Questions

1. Define the following terms connected with lift installations: (*a*) interval; (*b*) quality of service; (*c*) round-trip time.

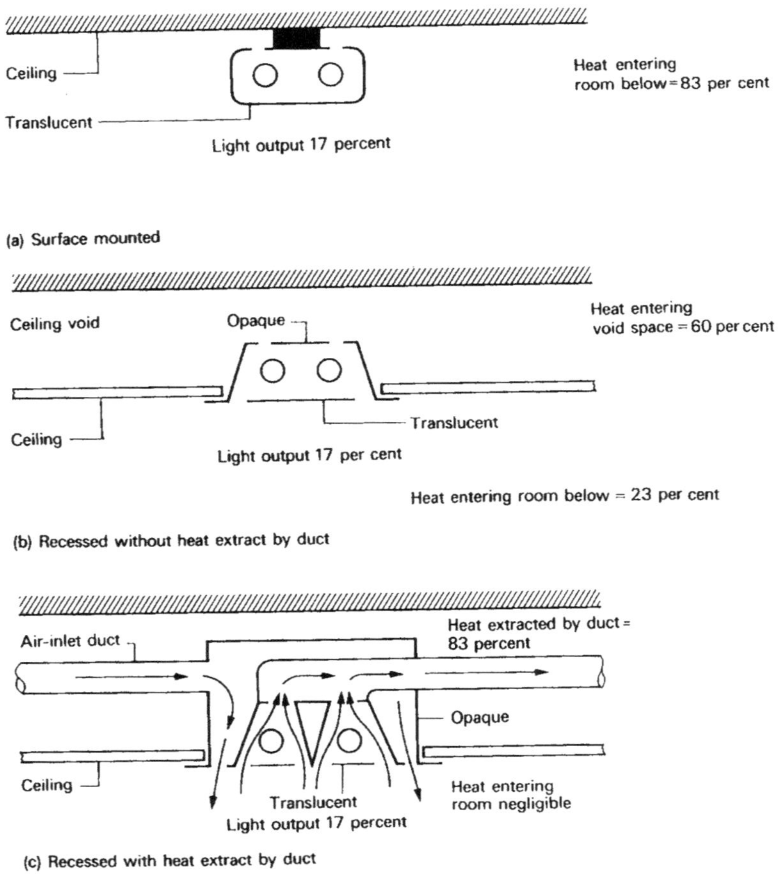

Fig.14.5 Energy balance in light fittings

2. A group of four lift cars each having a carrying capacity of 16 persons and a car speed of 1.5 m/s is specified to serve a ten-storey office having room heights of 2.5 m each. The net floor area above the ground level is to be 12 000 m^2 and a population density of 1 person per 10 m^2 of net floor area is to be used; and 17 per cent of this population will require transportation during the 5-minute peak period.
Calculate the round trip time for one lift car, the interval and capacity for the group and the quality of service provided.

Answers: (*a*) Round-trip time, 127.3 seconds
(*b*) Interval, 32 seconds
(*c*) Capacity for the group, 121 persons per 5 minutes
(*d*) Quality of service, acceptable

3. Explain the inverse-square law and the cosine law of illumination.

4. A spotlight has a light intensity along its beam of 18 000 candelas and is directed so that the beam is at right angles to a surface 8 m from the light source. Calculate the maximum illumination on the surface.

Answer: 281.25 lux

5. A diffusing spherical light fitting is suspended so that its centre is 2.5 m above the working plane. If the fitting emits 8000 lumens uniformly in all directions, calculate the intensity of illumination on the working plane.

Answer: 254.6 lux (approx.)

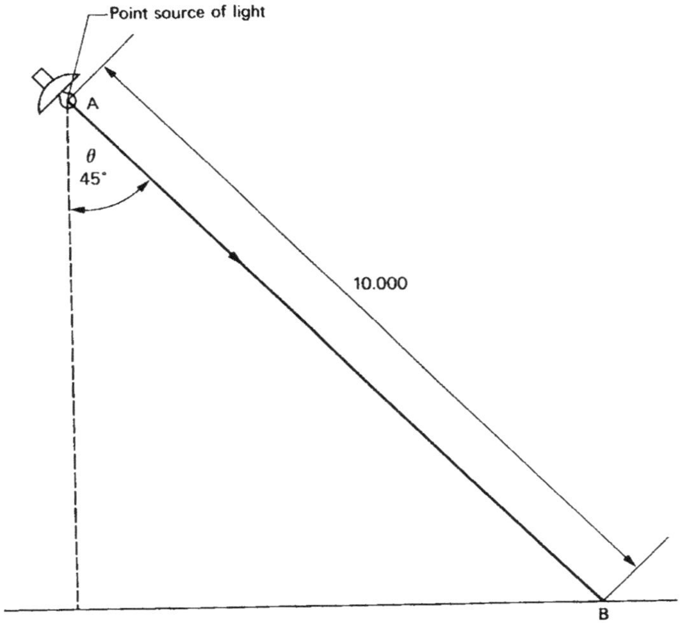

Fig 14.6

6. Figure 14.6 shows a light source at A directed on a horizontal surface at B. If the intensity of light at the source is 15 000 candelas, calculate the illumination on the horizontal surface.

Answer: 106 lux (approx.)

7. A general office of dimensions 8 m × 5 m × 3 m high is to be illuminated to a design level of 400 lux using 65 W fluorescent fittings having a BZ classification of 4 and an installed flux of 3500 lumens per fitting.
The working plane of the office is to be 850 mm above the floor so that the mounting height of the fittings is 2.15 m. If the utilisation and maintenance factors are 0.45 and 0.8 respectively, design a lighting scheme for the office.

Answers: 14 fittings are required providing a spacing of 2.7 m maximum centre to centre of fittings. Maximum distance from the centre of fittings to the wall is 1.35 m.

8. Define the following lighting terms: (*a*) utilisation factor; (*b*) maintenance factor; (*c*) lumen; (*d*) lux; (*e*) room index.

9. Outline the points to be observed in the design of electric lighting in accordance with the Illuminating Engineering Society Code.

Appendix

SI units (International System of units)

In 1971, the Council of Ministers of the European Community (EEC) decided to commit all members countries to amend their legislation in terms of SI units. The United Kingdom had already decided that SI units would become the primary system of measurement and legislation is established in several countries.

There are three classes of SI units, namely:

1. Base units
2. Supplementary units
3. Derived units

Base units

The SI system is based on seven units.

Quantity	Unit	Symbol
Length	metre	m
Mass	kilogram	kg
Time	second	s
Electric current	ampere	A
Thermodynamic temperature	kelvin	K
Luminous intensity	candela	cd
Amount of substance	mole	mol

Note: For ordinary temperature, the degree Celsius (°C) is used and the difference between two ordinary temperatures i.e. temperature interval degree Kelvin (K) is used.

Supplementary units

Quantity	Unit	Symbol
Plane angle	radian	rad
Solid angle	steradian	sr

The radian is the angle between two radii of a circle which cut off on the circumference an arc equal in length to the radius.

The steradian is an angle which, having its vertex in the centre of a sphere, cuts off an area of the surface of the sphere equal to that of a square having sides of length equal to the radius of the sphere.

Derived units

These are expressed algebraically in terms of base units and/or supplementary units. The main units are:

Quantity	Name of derived units	Symbol	Units involved
Frequency	hertz	Hz	1 Hz = 1 s^{-1} (1 cycle per second)
Force	newton	N	1 N = 1 kg m/s^2
Pressure and stress	pascal	Pa	1 Pa = 1 N/m^2
Work, energy and quantity of heat	joule	J	1 J = 1 N m
Power	watt	W	1 W = 1 J/s
Quantity of electricity	coulomb	C	1 C = 1 A s
Electrical potential, potential difference, electromotive force	volt	V	1 V = 1 W/A
Electric capacitance	farad	F	1 F = 1 A s/V
Electric resistance	ohm	Ω	1 Ω = 1 V/A
Inductance	henry	H	1 H = 1 V s/A
Luminous flux	lumen	lm	1 lm = 1 cd sr
Illuminance	lux	lx	1 lx = 1 lm/m^2

Multiples and sub-multiples of SI units

Factor	Prefix		
	Name	Symbol	
One billion (one million million)	10^{12}	tera	T
One thousand million	10^{9}	giga	G
One million	10^{6}	mega	M
One thousand	10^{3}	kilo	k
One hundred	10^{2}	hecto	h
Ten	10^{1}	deca	da
One tenth	10^{-1}	deci	d
One hundredth	10^{-2}	centi	c
One thousandth	10^{-3}	milli	m
One millionth	10^{-6}	micro	μ
One thousand millionth	10^{-9}	nano	n
One million millionth	10^{-12}	pico	p

Units for general use

Quantity	Unit	Symbol	Definition
Time	minute	min	1 min = 60 s
	hour	h	1 h = 60 min
	day	d	1 day = 24 h
Plane angle	degree	°	$1^\circ = (\pi/180)$ rad
	minute	′	$1' = (1/60)^\circ$
	second	″	$1'' = (1/60)'$
Volume	litre	l	1 l = 1 dm^3 (1000 l = 1 m^3)
Mass	tonne	t	1 t = 10^3 kg

Index

For Product Safety Concerns and Information please contact our EU
representative GPSR@taylorandfrancis.com
Taylor & Francis Verlag GmbH, Kaufingerstraße 24, 80331 München, Germany

www.ingramcontent.com/pod-product-compliance
Ingram Content Group UK Ltd.
Pitfield, Milton Keynes, MK11 3LW, UK
UKHW050059200926
475702UK00019B/2924

* 9 7 8 0 5 8 2 2 3 1 3 9 9 *